The Open University

SCIENCE MATTERS

Agriculture

prepared for the Course Team
by Pat Murphy

Science: a second level course

The S280 Course Team

Pam Berry (Text Processing)
Norman Cohen (Author)
Angela Colling (Author)
Michael Gillman (Author)
John Greenwood (Librarian)
Barbara Hodgson (Reader)
David Johnson (Author)
Carol Johnstone (Course Secretary)
Hilary MacQueen (Author)
Isla McTaggart (Course Manager)
Diane Mole (Designer)
Joanna Munnelly (Editor)
Pat Murphy (Author)
Ian Nuttall (Editor)
Pam Owen (Graphic Artist)
Gillian Riley (Editor)
Malcolm Scott (Author)
Sandy Smith (Author)
Margaret Swithenby (Editor)
Jeff Thomas (Course Team Chair and Author)
Kiki Warr (Author)
Bill Young (BBC Producer)
External Assessor: John Durant

Authorship of this book

Chapters 2–7 are based on major written contributions from experts outside the *Science Matters* Course Team. These contributions have been edited to varying extents to suit the purposes of the Course. Naturally, the Course Team accepts full responsibility for any errors or deficiencies introduced. We are very grateful to our consultants: Dr Philip John, University of Reading (Chapter 2); Dr Julian Wiseman, University of Nottingham (Chapter 3); Dr Tom Addiscott, Rothamsted Experimental Station (Chapter 4); Dr Anne Martis, an Open University tutor, who extensively rewrote and updated material originally produced by Dr Tom Chamberlain, University of Reading (Chapter 5); Professor H. F. van Emden, University of Reading (Chapter 6); and Dr Anne Martis, an Open University tutor (Chapter 7).

The second edition of this book incorporates an extensively rewritten Chapter 5 (BSE) and a replacement Chapter 7 (Integrated Farming Systems). The remaining chapters are unchanged from the first edition.

The Open University, Walton Hall, Milton Keynes, MK7 6AA.

First published 1993, Second edition 1997. Reprinted 1998

Edited, designed and typeset in the United Kingdom by the Open University.

Printed in the United Kingdom by The Burlington Press, Foxton, Cambridge CB2 6SW

ISBN 07492 81723

This text forms part of an Open University Second Level Course. If you would like a copy of *Studying with the Open University*, please write to the Course Reservations and Sales Centre, PO Box 724, The Open University, Walton Hall, Milton Keynes, MK7 6ZS. If you have not already enrolled on the Course and would like to buy this or other Open University material, please write to Open University Educational Enterprises Ltd, 12 Cofferidge Close, Stony Stratford, Milton Keynes, MK11 1BY, United Kingdom.

19388C/S280agri2.2

Contents

1	*Introduction*	*5*
2	*Oilseed rape*	*9*
2.1	Introduction	9
2.2	The erucic acid of rapeseed oil	11
2.3	The toxic glucosinolates of rapeseed meal	14
3	*Eggs and bacon*	*16*
3.1	Introduction	16
3.2	Egg yolk colour	16
3.3	Environmental influences on egg production	19
3.4	Nutrient provision in pig production	20
3.5	Qualitative aspects of pig-meat	23
Summary of Chapter 3		31
4	*The 'nitrate problem'*	*32*
4.1	'I only want my rights!'	32
4.2	Why worry about nitrate in water?	34
4.3	Where does nitrate come from?	37
4.4	Who controls nitrate on the farm?	40
4.5	What really happens to nitrogen fertilizer?	42
4.6	The Nitrate Sensitive Areas schemes	46
4.7	Who else contributes to the nitrate problem?	47
Summary of Chapter 4		48
5	*Bovine spongiform encephalopathy (BSE)*	*50*
5.1	Introduction	50
5.2	Cattle farming systems	52
5.3	Investigating the disease	53
5.4	What is the agent that causes BSE?	61
5.5	The risk of transmission of BSE to humans	63
Summary of Chapter 5		67
6	*Pesticides*	*69*
6.1	Introduction	69
6.2	The discovery and development of pesticides	69
6.3	The contents of the poison cupboard	71

6.4 Application of pesticides 76
6.5 The fate of pesticides 83
Summary of Chapter 6 87

7 Integrated farming systems *89*

7.1 Introduction 89
7.2 Economic and political change 90
7.3 Integrated farming techniques 92
7.4 An integrated farming systems approach 99
7.5 Future developments 102
Summary of Chapter 7 103

Further reading *104*

Skills *105*

Answers to questions *106*

Answers to activities *112*

Acknowledgements *130*

Index *131*

* Tape 1, Side 2 "Looking at data before starting this text.

Question 3. TMA 1 after studying all this book.

1 Introduction (1 hour)

It is certainly no exaggeration to claim that modern civilization is entirely dependent on agriculture. This is true in two senses. First, almost all our food, much of our textiles and a substantial amount of the other materials we use are produced by agricultural systems (particularly when these are defined to include forestry as well). Second, and just as significantly, civilization could not begin to develop beyond quite a rudimentary level until agriculture replaced hunting/gathering as the predominant means by which we fed ourselves.

An important aspect of this change was that agriculture, at least beyond slash-and-burn arable farming and nomadic-pastoralism, implies a relatively settled existence and hence the possibility of increasingly permanent buildings. In fact, quite apart from providing shelter, such buildings would be needed to store (for instance) harvested grain — both so that it remained edible until the next successful harvest and so that enough seed remained viable for sowing the next crop. Given the means of storing food and seed-corn, other possessions inevitably accumulated.

Another important aspect of the change must have been an overall increase in efficiency, in that the practice of agriculture provided more 'free time' after survival had been assured than had hunting/gathering. This time would not have been equally available to everybody. Instead, small, but increasing, numbers of people would have been released from the business of actually providing food to begin the development of art, written communication, philosophy, literature, mathematics, science, technology, and so on (in other words, most of what we call 'civilization').

Science, therefore, owes its origins to settled agriculture. It has also contributed massively to the development of agriculture, particularly since the 18th century when people like 'Turnip' Townsend started to experiment systematically with new agricultural practices and to breed improved varieties of crops and farm animals. The era of the professional expert inevitably followed. Extraordinary as this now seems, rubber as a crop is largely the single-handed achievement of one such expert, H. N. Ridley, Director of the Singapore Botanic Garden in the late 19th century. Nowadays nearly all agricultural science, and much agricultural education and training, is in the hands of teams of professionals.

But it is important to keep this in perspective. According to the plant breeder N. W. Simmonds, the total genetic change achieved by farmers since agriculture began was probably far greater than that achieved during the last 100–200 years by more systematic science-based breeding. Of the 127 major crops he analysed, Simmonds concluded that 11% were domesticated 9 000–7 000 years ago, 65% 7 000–2 000 years ago, 15% 2 000–250 years ago and only 9% during the past 250 years. It is assumed that the process of selecting the best plants from which to breed commenced soon after each species' initial domestication. This required both the means of, and an understanding of the necessity for, maintaining pure seed stocks. This may not have been too difficult in the case of the self-fertilizing cereals (such as wheat and barley) domesticated in the Near East, but was far from trivial in the case of cross-fertilizing maize domesticated by American Indians. The latter point emphasizes that agricultural innovations have been made all over the world. Again, Simmonds concluded that 32% of major crops were domesticated in the Near East and Europe, 30% in Central and East Asia, 15% in Africa and 23% in America.

Despite the historical dispersion of agricultural innovation in both time and space, the increasing influence of science (which can sometimes hardly be distinguished from technology) has both quickened the pace of change and concentrated its origins in the industrially and economically more powerful countries, particularly during the last 50 years. Along with a realization of the extent of these changes, and of the fact that not all of them have been universally beneficial, has come something of a reaction. Many people are coming to the conclusion that we ought to return to forms of agriculture that require less energy and fewer raw materials to be imported onto the farm and that place more reliance on the biological control of pests, etc. This call for less ecologically intrusive agriculture has, nevertheless, to be reconciled with the need to feed a human population that is probably about 1 000 times larger than when agriculture began—and which will inevitably continue to grow for some time to come.

It is clear, therefore, that urgent answers are required to some difficult questions. Should our aim be to return to an 'idyllic', pre-scientific era in which 'natural' and 'traditional' methods of farming predominate? Should we dismiss such a course of action as hopelessly unrealistic and encourage even further intensification of agriculture? Should we instead seek a 'middle way', in which science (which is as much concerned with understanding how the natural world operates as with inventing new means of conquering it) is directed towards the development of less ecologically damaging, but sufficiently productive, agricultural practices? The Course Team hope that, as you study this book, you will continually question the role, achievements and short-comings of science with respect to agriculture. We also hope that such active questioning on your part will help equip you to look equally critically at any issues related to agriculture and science that you come across outside the context of *Science Matters*.

Although this book has been given the title *Agriculture*, it does not attempt to cover comprehensively the role of science in agriculture—that task would be far too great. Instead, it takes six agricultural topics (represented by Chapters 2–7) in which science has had (and is having) a significant impact, and uses them to exemplify some aspects of the relationship between agriculture and science. The first five topics relate in various ways to what we have come to regard as 'conventional' agriculture, while the last focuses on what is probably best termed 'sustainable' agriculture. It should not be assumed that those involved in the production of the book condone all the agricultural practices discussed. We do recognize the importance of (for example) ethics in relation to intensive animal farming (Chapter 3), but have chosen to devote our efforts and your study time mainly towards the more explicitly scientific aspects of the topics covered. One of the criteria brought to bear on the selection of topics was the scope they provide for the practice and development of skills.

Before you begin your study of Chapter 2, we would like you to try Activity 1.1. This is intended to 'set the scene' by emphasizing the vast changes that have occurred in agriculture in the UK over the past half century or so. Some of these changes are attributable (at least in part) to scientific developments and some have provided (again, at least in part) the impetus for those developments.

Activity 1.1 *You should spend up to 15 minutes on this activity.*

Tables 1.1 and 1.2 present various agricultural statistics in the form of ratios comparing the situation in 1986 with that in 1936. A ratio of 1.0 (which is sometimes written 1 : 1) would imply that there had been no overall change between the two years. Ratios greater than 1.0 (>1.0) imply overall increases, and those less than 1.0 (<1.0) overall decreases, between 1936 and 1986.

The UK population was approximately 47 million in 1936 and approximately 57 million in 1986. Since 57 million/47 million = 1.2, the 1986 to 1936 ratio for population size is 1.2 (i.e. for every 1 person in 1936 there were 1.2 people in 1986, a 20% increase in the number to be fed, etc.).

Examine Tables 1.1 and 1.2 systematically; then *describe* the significant changes over the 50 year period and try to *explain* them briefly.

Table 1.1 The ratio of the quantities of some resources used in UK agriculture in 1986 compared with 1936.

Resource	Ratio 1986/1936
(a) resources intrinsic to the farm	
total area farmed	0.93
number of farmers	0.88
number of farm workers	0.37
(b) resources from outside the farm	
number of tractors	16
amount of nitrogen fertilizer	29

Table 1.2 The ratio of the tonnages of commodities produced in the UK in 1986 to those produced in 1936.

Commodity	Ratio 1986/1936
(a) grain crops harvested by combine	
wheat	9.3
barley	13
oats	0.25
beans	2.4
peas	11
oilseed rape	(not grown in 1936)
(b) root crops for human consumption	
potatoes	1.4
sugar beet	2.5
(c) crops for livestock	
turnips and swedes	0.32
mangolds and fodder beet	0.17
hay	0.56
silage	9.2
(d) animal products	
beef and veal	1.6
mutton and lamb	1.5
pig meat	2.5
poultry meat	12
liquid milk	1.5
butter	8.2
cheese	4.6
eggs	1.9
wool	1.2

An important factor contributing to the intensification of agriculture and the overall increases in production evident in Tables 1.1 and 1.2, was the political decision taken during the Second World War to make the UK more self-sufficient in food. This policy was pursued with great vigour and considerable success for decades afterwards. That the existence of 'grain mountains', etc. has prompted a recent re-appraisal of agricultural policy does not detract from the achievements of the farming community and others who have, after all, been implementing decisions ultimately taken by governments which most of us have had a part in electing.

Plate 2.1 Flowering oilseed rape in southern England.

Plate 3.1 Yolk colour fan. ▲

Plate 3.2 Slices of bacon showing differences in the amount of muscle fat. ▼

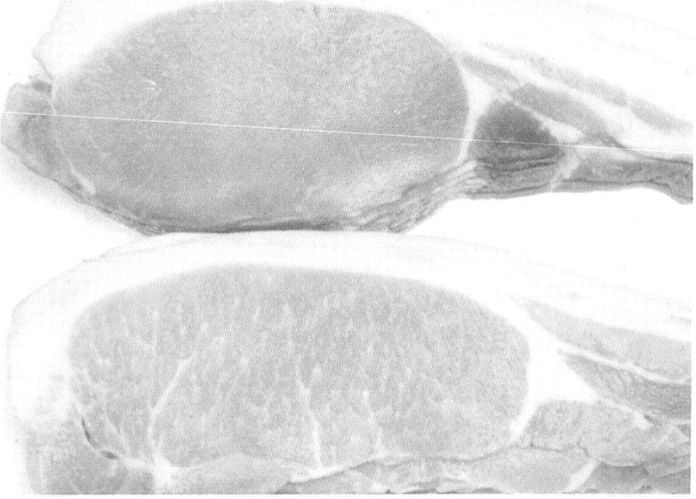

Plate 4.1 *Winter Landscape* by Rowland Hilder, a well-known 20th century British painter.

2 Oilseed rape (2 hrs)

① Wheat
② Barley
③ Oilseed Rape

2.1 Introduction

In early summer much of lowland Britain is dominated by a vivid yellow patchwork created by the flowering of the **oilseed rape** crop (Plate 2.1). Since the early 1970s this crop has risen from being an insignificant feature of our agricultural landscape until it is now the most widely grown arable crop after wheat and barley. Figure 2.1 shows for wheat, barley and oilseed rape (a) the area devoted to the crop and (b) the size of the annual harvest during the 1970s and 1980s. Figure 2.2 shows the oilseed rape data plotted on a scale more appropriate for that crop.

Activity 2.1 *You should spend up to 15 minutes on this activity.*

(a) For *wheat*, for *barley* and then for *oilseed rape*, make brief notes on any UK trends over the period 1970–1990 that are evident from Figures 2.1 and 2.2 in (i) area devoted to the crop, (ii) annual production and (iii) yield per unit area.

(b) What factors might have contributed to any trends you have identified in the yield per unit area for these crops?

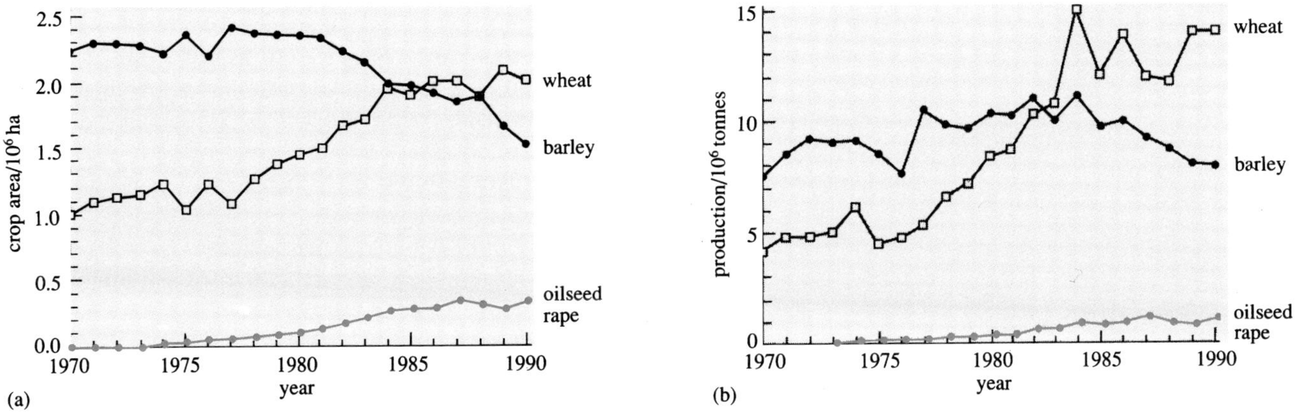

Figure 2.1 (a) Area devoted to the crop (in millions of hectares) and (b) annual production (in millions of tonnes) for wheat and barley in the UK, and oilseed rape in England, for 1970–1990.

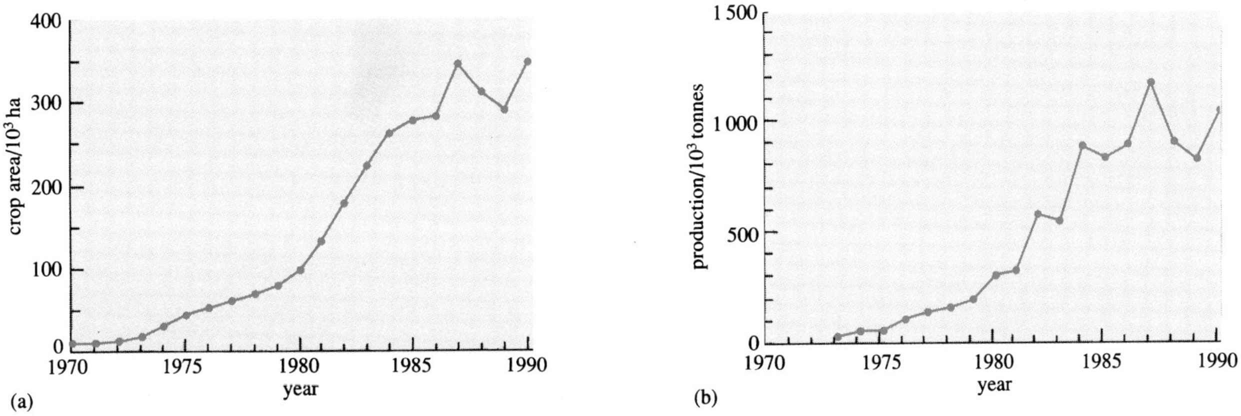

Figure 2.2 (a) Area devoted to the crop (in thousands of hectares) and (b) annual production (in thousands of tonnes) for oilseed rape in England for 1970–1990. (Production figures prior to 1973 are not available.)

What led to such a rapid transformation in the farming industry? The UK's entry (on 1 January 1973) into the EEC, with its policy of agricultural price support, was important. The increasing demand by consumers for vegetable oils helped. But oilseed rape could not have taken advantage of these factors if plant breeders had not been able to create varieties which simultaneously satisfied both the farmer and the oilseed industry. The plant breeder in turn was guided by the plant biochemist who was able to identify the qualities that should be aimed for. The biochemist's message to the breeder was to eliminate two undesirable chemicals, **erucic acid** and **glucosinolate**, from the old varieties (see Sections 2.2 and 2.3 respectively).

As its name suggests, oilseed rape is grown primarily for its oil, which constitutes about 40% by weight of the seed. It currently ranks fourth in the world as a source of oil after soya bean, palm and sunflower, having overtaken groundnut (peanut) in the 1980s. It is the only source of edible oil which can be grown successfully in the northern latitudes of Europe and it supplies one-third of the oil we use for cooking, salad oils and margarine. When the oil has been extracted by crushing the seed in a mill, there remains a residue called **rapeseed meal** (or **cake**). The high protein content (35–40%) and high nutritional quality of the meal makes it valuable as a feed for livestock, especially pigs and poultry. In particular, it has a good balance of the amino acids that these animals cannot themselves synthesize and which must therefore be present in their diet—the so-called **essential amino acids**. Although the protein is less valuable than the oil, it does make an important contribution to the economics of the crop.

Oilseed rape belongs to the Cruciferae, a family of flowering plants that also includes cabbages and kales, turnips, swedes, radishes, mustards and watercress. In fact, there are oilseed varieties of four closely related species of the genus *Brassica*: *B. campestris* (**turnip-rape**), *B. carinata* (**Ethiopian rape**), *B. juncea* (**mustard rape**) and *B. napus* (**rape**). Each of these species has a different **diploid number** of chromosomes, i.e. the number of chromosomes in the plants' non-reproductive (i.e. somatic) cells. For instance, Ethiopian rape has 34 chromosomes ($2n = 34$) arranged in 17 **homologous** (matching) **pairs**. Its **haploid number**, n, is therefore 17. It is thought that Ethiopian rape arose from a natural cross between *B. nigra* (black mustard, $2n = 16$) and *B. oleracea* (cabbage/kale, $2n = 18$). Similarly, mustard rape ($2n = 36$) is thought to have arisen from a cross between black mustard ($2n = 16$) and turnip-rape ($2n = 20$); and rape ($2n = 38$) from a cross between cabbage/kale ($2n = 18$) and turnip-rape ($2n = 20$). The relationship between these six species is shown in Figure 2.3.

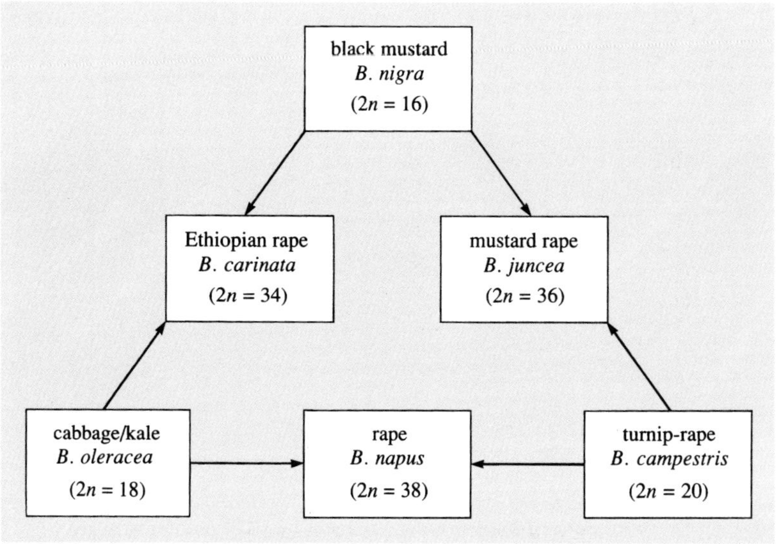

Figure 2.3 The relationship between six *Brassica* species.

Question 2.1 What pattern is evident in the diploid chromosome numbers of these six species? On the basis of this pattern, suggest how *fertile* hybrids could have arisen. It might help to consider first what normally happens to chromosomes during meiosis.

don't worry if can't answer.

Ethiopian rape is important only in its country of origin. Mustard rape is grown mainly in the Indian sub-continent and China; however it is also grown on a small scale in eastern England for use in the production of 'Colmans mustard' (it supplies the 'pungency', while white mustard (*Sinapis alba*) supplies the 'heat'). Rape and turnip-rape are the main oilseed species grown in Europe and Canada. There are **annual** (sown in the spring, harvested in the autumn) and **biennial** (sown in the autumn, harvested the following summer) varieties of both species. In general, autumn-sown (i.e. 'winter') varieties yield about 20% more than spring-sown varieties of the same species and rape yields about 20% more than turnip-rape sown at the same time of year.

▷ Other things being equal, which of these four crop types would a farmer prefer to grow?

▶ Winter rape, because it should yield more than spring rape, winter turnip-rape or (especially) spring turnip-rape.

Winter turnip-rape is somewhat more winter-hardy than is winter rape; therefore, in locations where the winters are slightly too severe for winter rape, winter turnip-rape will be grown instead. Where the winters are even more severe, spring rape will be grown. Finally, since turnip-rape is somewhat faster-maturing than rape, spring turnip-rape is grown where the season is too short for spring rape. While winter rape varieties predominate in the UK and most of western Europe, in Canada approximately equal areas are devoted to spring rape and spring turnip-rape.

Winter rape is sown in early autumn, when the ground is relatively dry and easily worked. Following germination, the seedlings grow until it becomes too cold. Growth resumes early in the spring and the plants become quite large, producing correspondingly high yields of seed and oil. The crop is harvested during the summer using the same basic machinery employed for cereals. Generally, fields sown with winter rape one year are sown with winter wheat the next.

2.2 The erucic acid of rapeseed oil

Plant oils, for which rape is primarily grown, are known as **triglycerides** because they are esters of **glycerol** and three **fatty acid** molecules (which may be of the same kind or different).

Question 2.2 Write a balanced chemical equation for the formation of a molecule of glyceryl tripalmitate from one molecule of glycerol and three molecules of palmitic acid.

don't worry if can't answer.

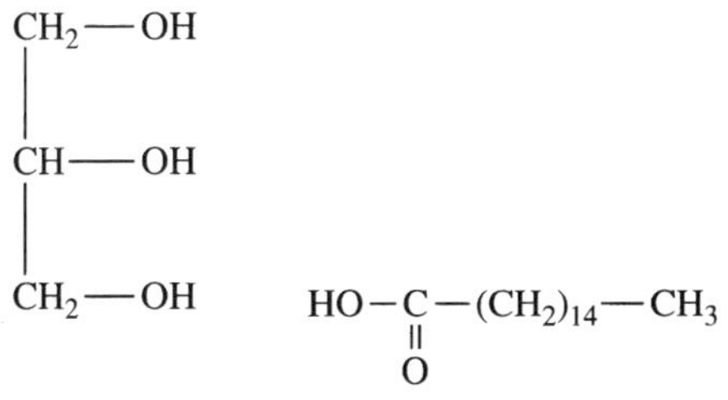

glycerol palmitic acid

Figure 2.4 shows the structure of a slightly more complex triglyceride, containing two oleic acid residues and one linoleic acid residue.

$$CH_2-O-\overset{O}{\underset{\|}{C}}-(CH_2)_7-CH=CH-(CH_2)_7-CH_3 \quad \text{oleic}$$

$$CH-O-\overset{O}{\underset{\|}{C}}-(CH_2)_6-(CH_2CH=CH)_2-(CH_2)_4-CH_3 \quad \text{linoleic}$$

$$CH_2-O-\overset{O}{\underset{\|}{C}}-(CH_2)_7-CH=CH-(CH_2)_7-CH_3 \quad \text{oleic}$$

glycerol residue | fatty acid residues

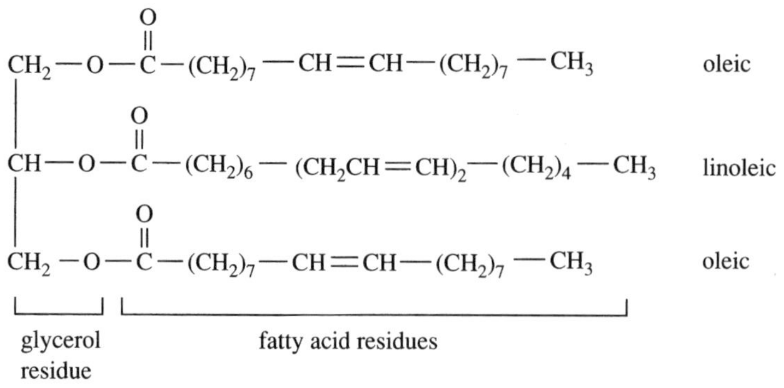

Figure 2.4 The structure of a triglyceride; in this example the glycerol is esterified with two oleic acid molecules and one linoleic acid molecule.

As Figure 2.4 implies, fatty acids are composed of chains of carbon atoms with varying numbers of double bonds. Fatty acids with one or more carbon–carbon double bonds (C=C) are termed **unsaturated**; those without any such double bonds are termed **saturated**. A convenient 'short-hand' for fatty acids is to write them in the form $Cx:y$ (where x represents the number of carbon atoms and y the number of double bonds).

Question 2.3 What are the short-hand forms of palmitic, oleic and linoleic acids?

In most edible oils, the main fatty acids combined with glycerol are palmitic, oleic and linoleic. However, the older varieties of oilseed rape contained high concentrations of erucic (and eicosenic) acid (Table 2.1).

Table 2.1 The fatty acid composition of some vegetable oils, expressed as a percentage of the total fatty acids.

Fatty acid	Rapeseed oil (from older varieties)	Soya bean oil	Sunflower oil	Corn oil
palmitic acid	3–5	10–11	5–7	10–12
oleic acid	18–27	23–26	20–23	26–28
linoleic acid	14–18	50–54	64–68	55–58
linolenic acid	8–9	7–9	trace	trace
erucic acid*	37–59	—	—	—

* In fact, erucic acid plus a small amount of eicosenic acid.

Linoleic acid (vitamin F) is a component of human cell membranes and is also required in our metabolism; since neither we nor other mammals are capable of synthesizing it, a source is required in our diet. Erucic acid, however, does not normally participate in animal cell structure and metabolism; it is detected in animals only when it has been consumed in large quantities. When laboratory animals were fed large amounts of rapeseed oil in the 1960s, a variety of deleterious effects were observed: retarded growth, disturbed reproduction and heart lesions. Although such effects have never been observed with humans, the results of these laboratory tests were regarded as a warning and it was decided that the erucic acid content of rapeseed should be reduced.

Figure 2.5 is a simplification of the biosynthetic pathways in oilseed rape leading from palmitic to other fatty acids.

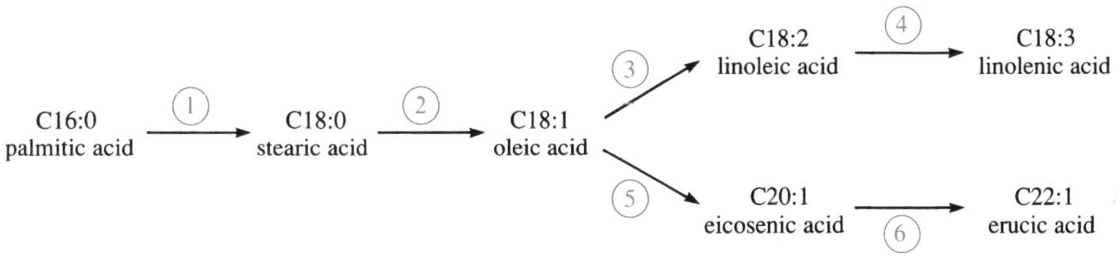

Figure 2.5 Simplified biosynthetic pathways in oilseed rape leading from palmitic to other fatty acids.

▷ What overall trends are evident in the pathways leading from palmitic acid to linolenic acid and from palmitic acid to erucic acid?

▶ The numbers of carbon atoms increase in steps of two (from 16 in palmitic acid to 18 in linolenic acid and 22 in erucic acid). There is also an increase in the numbers of carbon–carbon double bonds (from 0 in palmitic acid to 1 in erucic acid and 3 in linolenic acid).

Each step in a biosynthetic pathway (represented by a numbered arrow in Figure 2.5) is catalysed by one or more enzymes. If an enzyme involved in a particular step is missing (or deficient in some way), the pathway becomes blocked at that step and products 'downstream' of the blockage will not be synthesized (at least not by *that* pathway—other pathways *may* be available).

Question 2.4 If you wanted to reduce the amount of erucic acid in oilseed rape, but without loss of many other unsaturated fatty acids, which would be the best step to have blocked?

Plant breeders found that a Canadian variety of spring rape had a low content of erucic acid. By a series of crosses they were able to incorporate this character into winter rape varieties grown in Europe. However, the first of these new low-erucic-acid varieties of winter rape produced significantly less oil than the winter varieties from which they were derived. The explanation for this turned out to be remarkably simple. Erucic acid (C22:1) molecules combined with glycerol in triglycerides in the original varieties were replaced by oleic acid (C18:1) or linoleic acid (C18:2) molecules in the new low-erucic-acid varieties. Each such replacement represented a shortening of the fatty acid chain by four CH_2 units, which was translated into a measurable reduction in the mass of oil. Later breeding increased the oil content by increasing the number of molecules of triglyceride the plants produced, so that the initial yield penalty of the low-erucic-acid varieties disappeared.

The introduction of the new varieties was encouraged by the application of EEC regulations in the 1970s which limited the erucic acid content in vegetable oils for human consumption and imposed a maximum erucic acid standard for rapeseed sold into intervention (the system in which minimum prices are guaranteed to growers). By 1979 the EEC Commission had reduced the permissible quantity of erucic acid in vegetable oils to 5%. Low-erucic-acid varieties are now defined as those containing less than 2% erucic acid as a proportion of the measured fatty acids.

All the varieties now grown in the UK are low in erucic acid. Even so, the erucic acid content of each load of seed is carefully assayed before the oil is extracted at the mill. Occasionally a load sent for crushing is found to have a high erucic acid content. While no-one is sure how this arises, one possibility is that some of the seed that was sown resulted from cross-pollination with wild *Brassica* species with a high erucic acid content.

In fact, two distinct types of low-erucic-acid rapeseed oil are required for human consumption—one type is used in margarine production and the other for cooking and frying. These differ in the composition of their other fatty acids, and plant breeders strive to produce varieties which are improvements on their predecessors in this respect. The content of linolenic acid should be minimized in both types because this unsaturated acid is unstable and easily oxidized to give unpleasant-smelling substances. In margarine, it is desirable to have a high percentage of linoleic acid, both because it is an essential nutrient (vitamin F) and because it is thought to reduce the amount of cholesterol in the blood. For instance, sunflower oil (Table 2.1) has a much higher linoleic acid content (the *polyunsaturates* of 'Flora' margarine) than rapeseed oil and enjoys a commercial advantage as a result. However, since margarine has an undesirable 'sandy' texture when it contains relatively large crystals (such as are produced by unsaturated fatty acids with chain lengths of 18 carbon atoms), one of the current aims of rape breeders is to increase the percentage of palmitic acid (C16:0)—even though it is saturated! In varieties bred for frying oil, as well as reducing the percentage of readily oxidized linolenic acid, breeders aim to increase the percentage of oleic acid because it resists thermal breakdown.

Rather surprisingly, having succeeded in virtually eliminating erucic acid from rapeseed oil, plant breeders are now trying to develop varieties with high levels of erucic acid! The explanation is that there remains a demand from industry for such oils. Although they have a variety of technical uses in the chemical and steel industries, one application is particularly notable. Oils containing erucic acid are very water-repellent and have a good 'cling' to metal surfaces that are washed with water. This property made them useful lubricants for steam engines. Indeed, it was to provide a lubricant for marine engines that the cultivation of oilseed rape was begun in Canada in the 1940s. However, breeders have found an upper limit of 66% to the erucic acid content in rapeseed oil. This appears to be imposed by the restriction of erucic acid to the two outer positions in the triglyceride, with the middle position always being occupied by another fatty acid (see Figure 2.4).

▷ Biochemists have discovered that the oil in the seeds of the common garden nasturtium (*Tropaeolum majus*) contain 78% erucic acid. What does this imply?

▶ That, in this species, erucic acid must occupy *all three* positions in *some* of the triglyceride molecules in the nasturtium seed oil.

It is possible, therefore, that the barrier to the entry of erucic acid to the middle position may not be insuperable in oilseed rape. Although it is impossible to cross nasturtium with oilseed rape by conventional plant breeding methods, genetic engineers might soon be able to isolate the gene for the enzyme responsible for the insertion of erucic acid into the middle position and incorporate it into the genotype of oilseed rape, thereby overcoming the 66% limit.

2.3 The toxic glucosinolates of rapeseed meal

Rapeseed meal contains a high proportion of protein with a good balance of amino acids. Nevertheless, for many years its value as a feedstuff for livestock was limited, and it could be incorporated into animal feed mixtures only at low rates, because of the presence of glucosinolates. These sulphur-containing compounds are broken down (hydrolysed) by the enzyme myrosinase, which is released from crushed seed, to give bitter-tasting and toxic products. When the enzyme and its glucosinolate substrate come into contact, hydrolysis is rapid (Figure 2.6). The two main glucosinolates of rape seeds are gluconapin (in which the R group is but-3-enyl, $CH_2{=}CHCH_2CH_2{-}$) and progoitrin (in which the R group is 2-hydroxy but-3-enyl, $CH_2{=}CHCH(OH)CH_2{-}$).

Glucosinolates are a characteristic feature of the family Cruciferae; they are, for instance, responsible for the pungent taste of mustard and horseradish. However, not only do they reduce the palatability of rapeseed meal, they also have other undesirable

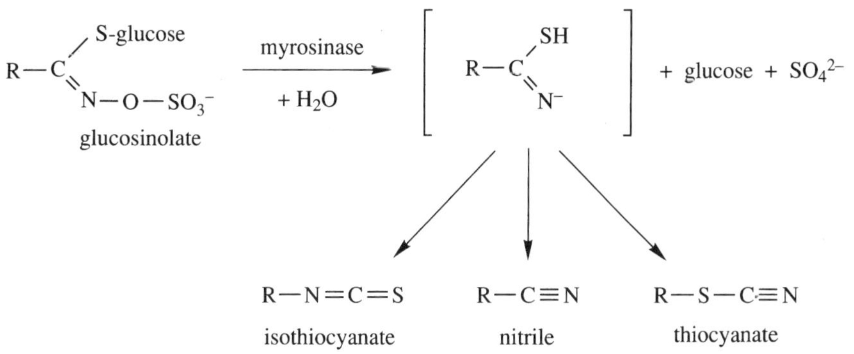

Figure 2.6 The hydrolysis of glucosinolates by myrosinase. (As is usually the case in biochemistry, each arrow represents *several* distinct chemical reactions; square brackets are used to denote an intermediate product/substrate in the hydrolysis.)

effects. Goitre is caused by the thiocyanate (Figure 2.6) released from progoitrin, which prevents iodine uptake into the thyroid gland. Additionally, the product of progoitrin hydrolysis interferes with synthesis of the hormone, thyroxin. All glucosinolates produce nitriles on hydrolysis (Figure 2.6), which are toxic to the liver and kidneys.

The solution to the glucosinolate problem became available to breeders in 1968, when it was discovered that a Polish variety of spring rape contained one-tenth of the normal levels of glucosinolates. The three genes responsible for the low levels of glucosinolates have now been bred into high-yielding winter varieties, which are also low in erucic acid. Tests have shown that the meal from these new varieties can safely be given to animals at higher rates than was possible with the high-glucosinolate varieties. In order to encourage the cultivation of the low-glucosinolate varieties, current EC regulations limit the glucosinolate content of seeds. Those with higher levels do not qualify for a subsidy. The varieties now grown are thus known as *double-low* varieties: low in erucic acid and low in glucosinolates.

That, however, is not the end of the story as far as the glucosinolates are concerned. The vegetative parts of the new double-low varieties can also have lower glucosinolate levels, although the reduction is far less marked than in the seeds. Glucosinolates and their hydrolysis products are thought to serve as part of the plant's defence against attack by insect pests and microbial diseases. They deter insect feeding as effectively as their pungency in feed deterred livestock. However, for those insects that have adapted to become specialized feeders on the Cruciferae (for example, the cabbage butterfly (*Pieris brassicae*) and the cabbage aphid (*Brevicoryne brassicae*)), these compounds can serve as attractants, and actually *stimulate* feeding and egg-laying. The new low-glucosinolate varieties may therefore be less prone to attack by these particular insects. Does this mean that the new varieties are inherently more susceptible to attack from pests and diseases which previously avoided oilseed rape, so that more pesticide has to be applied? Or will less chemical protection be needed because the new varieties will be attacked less by their traditional enemies, who will no longer be able to recognize the plants? It is not easy to predict the answers to these questions. The glucosinolates that have been reduced in the new varieties are only a few among the many chemicals that determine a plant's relationship to insect pests and microbial pathogens. The levels of these chemicals change as the plant grows and develops, and the levels can also alter in response to environmental conditions. Unravelling this complex relationship will be part of the continuing process of improving oilseed rape—and other crops—through scientific plant breeding.

Activity 2.2 *You should spend up to 20 minutes on this activity.*

Summarize the main points in this chapter as a numbered list. You will probably need to review the chapter first. However, you should be able to do this fairly quickly—especially if you highlighted key words and phrases during the first reading.

3 Eggs and bacon (3hrs)

3.1 Introduction

In spite of periodic 'food scares' and occasional political pronouncements, egg consumption in Great Britain remains high—of the order of 30 million per day. In addition to being consumed whole (boiled, fried, poached, scrambled and as omelettes), eggs are used in a wide variety of human confectionary foods (e.g. cakes and biscuits), as well as in pies and quiches. Their physico-chemical properties also make them useful constituents of the emulsions which are the basis of mayonnaise, etc. Eggs are an important and very nutritious part of the human diet—which is not surprising, for they have to meet all the nutrient requirements of the developing chick.

Consumption of pig-meat in Great Britain is around 21 kg per person per year (out of a total meat intake of approximately 55 kg). Compared to eggs, it is less easy to disguise the presence of pig-meat in food. Nevertheless, pig products are diverse and include fresh pork, cured items (e.g. bacon and ham), processed commodities (e.g. pies, sausages and salami) and fat (which may be used as lard in cooking or incorporated in products containing non-milk fat).

Eggs and pig-meat are **outputs** from just two of the many **animal production systems** maintained by the livestock industry. Looked at another way, they represent the end-products of the biological processes of ovulation and growth respectively. The food production industry endeavours to optimize these outputs and this can best be achieved if the basic science of these biological processes is understood, preferably down to the molecular level.

The more obvious **inputs** to the production systems are the animals themselves and the nutrients with which they are provided. However, other factors can more loosely be regarded as inputs, e.g. the environmental conditions under which the animals are kept, their disease status, the disposal of their waste products and welfare considerations. This chapter will focus on two of these inputs—nutrient provision and environmental conditions—and how they influence both the quantity and quality of the output.

3.2 Egg yolk colour

Crack an egg and observe the colour of its yolk. It will probably be a rich golden yellow. This is pleasing to the eye and you may well associate the colour with the egg being 'good for you'. Where, though, does the colour come from? Is it of fundamental importance to the egg's nutritional value? Why is this particular colour associated with 'wholesomeness'?

Pigmenting agents are distributed widely in nature and are responsible for the incredible variety of colours we see about us. As far as food is concerned, visual appeal has always been of particular importance and it is difficult to think of any foodstuff in which colour is not a significant factor contributing to its appeal. However, perceptions do change. White flour was originally popular because it apparently contained none of the impurities that unscrupulous millers and bakers were suspected of adding to increase weight at little extra cost. Nowadays, brown 'wholemeal' flour is in vogue because it is associated with 'health' and 'vitality'—again allowing scope for the addition of items which have never been near a wheat grain.

Food items that are of a bland neutral colour, whilst still being perfectly nutritious, are likely to be discriminated against in the market-place (an argument that could probably be extended to the packaging in which they are contained). Think for a moment about how you would respond to scrambled eggs whose colour was pale and insipid—and why you might respond in this way.

The chicken is unable to synthesize the pigmenting agents responsible for egg yolk colour. In the days before egg production became intensified, hens were free to roam around and scavenge for food. They would thus have had access to a wide range of vegetable matter containing various pigmenting agents. Some of these agents would ultimately end up in the yolk, which would thus be coloured accordingly. Since the actual colour of the yolk and its intensity would have been influenced by what the hens had been eating, the colours produced would certainly have been far from uniform. Indeed, the colours may sometimes have been rather unexpected. Maggots used for freshwater fishing are coloured deep red to attract fish. Cases have been reported of surplus maggots (which are extremely nutritious and well-liked by poultry) being fed to hens—which then oblige by producing eggs with crimson yolks! Would you want to eat one of these, probably quite innocuous, eggs?

These days, the colour of many foods appears to be associated not just with its visual appeal but also with its perceived degree of 'naturalness' and 'wholesomeness'. Despite the potentially wide variability in yolk colour as a consequence of the pigments consumed by 'free-range' hens, many consumers are probably convinced that a rich golden yellow/orange colour is 'best'. Although there is some evidence which suggests that egg yolk pigments *may* have an important biochemical role in humans, eggs of a particular yolk colour are certainly not produced for reasons of either nutrition or taste. It is up to the poultry industry to ensure that demand for eggs with the favoured yolk colour is fulfilled.

3.2.1 Defining yolk colour

In order to produce eggs of a particular yolk colour consistently, it is necessary to be able to define that colour. Although the colour can be defined *descriptively* (e.g. 'rich golden yellow/orange'), it is far better to *quantify* it in some way. The quantitative definition *could* refer to the occurrence of specific wavelengths within the visible spectrum of light reflected by the yolk. In practice, however, the information is represented in a 'yolk colour fan' (Plate 3.1) which consists of a *scale of colours* covering the range of values likely to be found in egg yolks.

Although somewhat subjective (as, indeed, are most evaluations of 'product quality'), assessing yolk colour with the fan is based on sound theoretical principles and does allow a useful grading system to be adopted. The following activity gives you an opportunity to try this for yourself.

Activity 3.1 *You should spend up to 45 minutes on this activity (not including time spent collecting data).*

(a) If eggs are used in your household, compare the yolk colours of the eggs used over (say) the period you are studying this book with the 15 colours of the yolk colour fan shown in Plate 3.1. (The colours are numbered 1–15 clockwise from palest to darkest.) Represent your data in the form of a bar chart and draw what conclusions you can from it. For instance, is there any evidence that the source of the eggs (e.g. large supermarket, farm shop) makes a difference?

(b) Show Plate 3.1 to a sample of people (preferably at least 30) and ask them which of the 15 colours is their *preferred* egg yolk colour. Again, represent your data as a bar chart and draw what conclusions you can from it. Is there any evidence of preference differences between males and females, between adults and children, or in relation to ethnic or cultural background? If so, it might be interesting to investigate further.

(c) What conclusions can you draw from comparison of the bar charts produced for parts (a) and (b)?

(d) Of course, conclusions are only as good as the data upon which they are based. In turn, the value of the data depends on how they are collected. What reservations do you have about the way we have suggested you collect the data for this activity?

3.2.2 Optimizing yolk colour

The 'optimum' yolk colour having been defined largely by the major food retailers, it now remains for the hens to be fed appropriately so that the required colour is produced consistently. Intensively reared hens are entirely dependent upon what has been provided for them to eat. Clearly these so-called laying diets must supply all the nutrients necessary to ensure optimum output of eggs. However, a nutritionally acceptable laying diet may not necessarily contain appropriate levels of the specific pigmenting agents needed to produce eggs of the required uniform colour. In any case, naturally occurring pigmenting agents are not particularly stable. Reliance on such natural agents as happen to occur in laying diets would almost certainly result in a level of variability in egg yolk colour that would be unacceptable to the major retailers (and possibly also to consumers).

Biochemical research has identified the natural pigmenting agents that are responsible for egg yolk colour and these compounds can now be synthesized in the laboratory.

▷ What advantages might the use of these synthesized compounds have over *exactly the same* compounds occurring naturally?

▶ Because the synthesized compounds can be more easily purified, they can be fed to hens at more accurately known concentrations than the same compounds produced naturally. The hens' response to them is thus likely to be more predictable.

So, the addition to the diet of precise amounts of so-called **'nature-identical' synthetic compounds** results in eggs having a highly predictable yolk colour. In other words, products of a standard quality can more readily be assured.

It is also possible to synthesize compounds which, whilst being *very similar* to naturally occurring pigmenting agents, are not quite *identical* to them and so are referred to as being **'nature-related' synthetic compounds**. Useful 'nature-related' compounds have all the desirable characteristics of 'nature-identical' compounds, as well as certain advantages over them—such as producing the desired effect at lower concentrations or being chemically more stable.

There is currently a definite adverse reaction among certain sections of the 'general public' to the use of synthetic chemical 'additives' in livestock diets. The use of both 'nature-identical' pigmenting agents (which are chemically indistinguishable from naturally occurring agents) and 'nature-related' agents (which are remarkably similar to them) have been criticized. In response to this reaction, interest has re-focused on naturally occurring sources of pigmenting agents (concentrated from plants such as marigold and paprika), even though their use to control egg yolk colour is somewhat less efficient and sometimes leads to unpredictable results.

3.2.3 Conclusions on egg yolk colour

The story of egg yolk colour is an interesting illustration of the general problem of meeting quality requirements for food products. The quality criterion in question is itself of little nutritional or even **organoleptic** (i.e. the complex of sensations that includes the smell and taste of a food) significance—the key feature is the visual impression created by the product. In the case of eggs, studies have revealed precisely how the quality of the product changes in response to changes in the dietary inputs. Indeed, the chemical compounds responsible have been identified, purified, stabilized and even synthesized. The industry has learnt how to incorporate them into poultry feed so that a highly predictable and consistent product results. Nevertheless, consumer reaction against the use of 'unnatural' synthetic chemicals has led to the use of concentrated natural sources which are not nearly as 'efficient'.

For both the producer and the consumer the current dilemma is rather philosophical in nature, revolving around the question: 'When is a chemical natural and when is it synthetic?' In terms of the response of (in this case) the hen, the distinction is quite spurious. However, the dilemma is quite perplexing for the retailer. Having attempted to convince consumers that (say) eggs with a particular uniform yolk colour represent the 'best buy'—and having created a market for a certain product quality (presumably with a view to maximizing its market share for that product)—retailers are now encountering resistance to the demand being satisfied by the most efficient means available. Perhaps they will have to attempt to change the demand again!

3.3 Environmental influences on egg production

Although many minor environmental factors affect egg production, by far the most important is **photoperiod** (i.e. the number of hours of 'daylight' experienced by the hens each 24 hours). The influence of photoperiod on egg production represents a good example of the physiological response of an animal to external stimuli.

Egg laying is the end-point of *ovulation* (i.e. the release of the egg, or *ovum*, from the ovary), which is a key stage in reproduction. After being shed from the ovary, an ovum travels down the oviduct during which time it accumulates a variety of tissues including, just before laying, its shell. Ovulation is under hormonal control and, in birds, the stimulus for its onset is usually increasing day length. Outside the tropics, this is of fundamental significance to wild birds in that it ensures that most species lay their eggs in the spring (which is presumed to be the most advantageous time of the year for reproduction). In the wild, therefore, sexual maturity would not normally be reached until the bird was at least a year old. However, under conditions of intensive egg production, this delay would be regarded as uneconomic since the birds would be unproductive for a considerable period of time whilst continuing to utilize resources (food, space, heat, etc.).

Since sexual maturity is controlled by photoperiod, it should be possible to induce birds to commence egg laying at a much younger age than they would in the wild by keeping them in lightproof houses and exposing them to varying 'day lengths' through the use of artificial lights. This is precisely what happens in intensive commercial egg production. Figure 3.1 presents a typical photoperiod regime to which hens kept for egg production are subjected from the time they hatch ('week 0') onwards. Under such a regime, the reproductive tract develops so that egg laying normally commences at between 22 and 25 weeks of age.

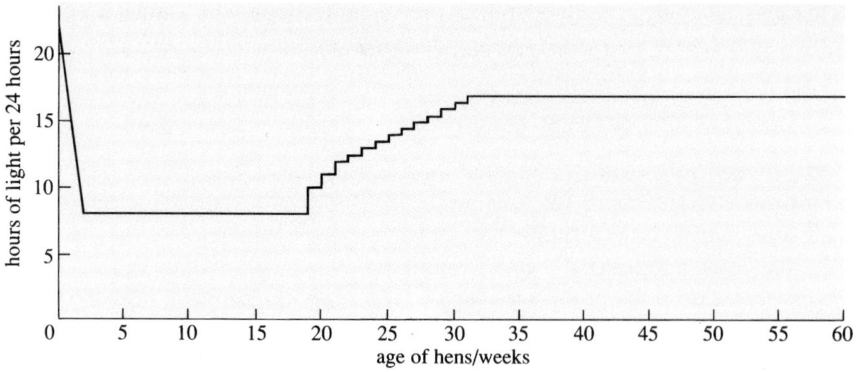

Figure 3.1 Typical photoperiod regime used in commercial egg production.

Question 3.1 Describe the photoperiod regime shown in Figure 3.1, relating it to what you know of the birds' response to natural variation in day length.

Alteration of the photoperiod in this way provides an excellent example of how an understanding of the underlying biological processes makes it possible to arrange conditions so that the efficiency of these processes is improved (at least from the point of view of humans).

3.4 Nutrient provision in pig production

Two hundred years ago, John Mills wrote that:

> *Of all the quadrupeds that we know, the hog appears to be the foulest, the most brutish and the most apt to commit waste wherever it goes... it devours indiscriminately whatever comes its way.*

This rather intemperate criticism does, in fact, encapsulate all that is valuable about the pig. As a true omnivore, it can thrive on waste products and has done so for centuries. Accordingly, it could supply meat to a large sector of the population at little cost. During the early Middle Ages, when hunger was a real problem, pig-meat was often the only source of many essential nutrients to the vast majority of the population and it has been claimed that the pig made a significant contribution to the development of medieval society in Europe. Even within living memory, its omnivorous habit was put to good use during the siege economy of the Second World War. Contemporary posters encouraged the population to save all their scraps for pig feeding.

Like all other biological organisms (including humans), pigs grow in response to the input of nutrients. It matters little to the pig whether these are supplied in the form of scraps or in the sophisticated concentrated diets of the modern intensive farm. What determines the rate and efficiency of growth is how closely nutrient provision matches the animals' requirements. Over the years, these requirements have been relatively precisely estimated, both qualitatively and quantitatively.

Pig production is concerned predominantly with the growth of muscle, the most important constituent of which is protein. Protein synthesis requires the presence of a number of individual amino acids. Those which the pig itself cannot synthesize have to be present in its diet (essential amino acids). Although it is convenient to estimate nutrient requirements in terms of protein, it is in fact the supply of these essential amino acids that is important. These and other essential nutrients are presented in

Table 3.1. It is interesting to note that the list for humans is very similar—although, as a result of systematic experimentation, the quantitative requirements of the pig are somewhat better known.

Table 3.1 Some of the essential nutrients required by pigs for growth.

Amino acids	Fatty acids	Vitamins	Major minerals	Minor minerals
lysine	linoleic acid	A	calcium	iron
methionine		B complex	phosphorus	copper
threonine		D	magnesium	zinc
tryptophan		E	potassium	manganese
		K	sodium	cobalt
				selenium

▷ Apart from these essential nutrients (and water), what else do pigs require in their diet?

▶ They must have a source of dietary energy to fuel all the metabolic processes associated with maintenance (primarily the need to maintain body temperature) and tissue synthesis (for example, linked to growth and repair).

Strictly speaking, energy is not a nutrient. It is supplied from the catabolism of other dietary commodities, referred to as 'energy-yielding'—principally carbohydrates and fats (protein is used to an extent, but this is 'wasteful' of valuable component amino acids). Conventionally for pigs, the 'energy-yielding potential' of the diet is referred to as 'digestible energy'. Carbohydrates, fats and proteins are also used as 'building blocks' in the synthesis of body tissues, for example dietary fat is often directly deposited into the carcass fat of animals.

All these requirements are provided in a diet composed of a number of individual ingredients. For instance, 1.5 kg per day of the diet presented in Table 3.2 would supply a young pig with all its food needs (including 22 MJ of energy, since the concentration of 'digestible energy' in the diet is 14.67 MJ kg^{-1}). Since pigs tend not to chew their food much, the ingredients have to be ground up and the whole diet presented in the form of a mash.

Table 3.2 Composition of a diet for young pigs in terms of ingredients.

Ingredients	Percentage of diet by weight
barley	35
wheat	24
fat premix	10
full fat soya	12.5
soya bean meal	10
fish meal	5
lysine	0.63
threonine	0.14
di-calcium phosphate	1.30
sodium chloride	0.18
supplement*	1.25

* Contains vitamins and minor minerals.

This diet can also be considered in terms of the nutrients it contains (Table 3.3).

Table 3.3 Composition of a diet for young pigs in terms of nutrients.*

Nutrients	Percentage of diet by weight
protein	20
oil	8.6
fibre	4.0
lysine	1.4
methionine	0.33
threonine	0.85
tryptophan	0.24
calcium	1.2
phosphorus	0.77
sodium chloride	0.6
linoleic acid	2.2

* Note that the diet would also contain energy-yielding constituents which are not regarded as nutrients.

Question 3.2 Which of the ingredients listed in Table 3.2 are also listed as nutrients in Table 3.3? Compare the percentages of the diet by weight that they represent as ingredients and as nutrients. How can any differences be explained?

As mentioned above, the nutrient requirements of pigs and humans are remarkably similar—in fact, the only major difference is that pigs do not appear to need vitamin C in appreciable amounts. Thus, the diet in Table 3.2, supplemented with an orange, would be quite satisfactory for us and its widespread use would go a considerable way towards eradicating world hunger. How would you feel about being presented with it every day? Your probable reaction illustrates the extent to which, for humans, food consumption has become a social and pleasurable activity rather than simply the means by which nutrient and energy requirements are met.

Any animal diet represents an attempt to find an optimum solution to a problem. The problem is how to combine ingredients of known nutritive composition into a diet of particular nutritive value (decided upon as a result of experiments) at the lowest cost. Nowadays, such diets are formulated using computer-based mathematical techniques.

The utilization by the pig of any nutrient requires its digestion, absorption and subsequent metabolism. Nutrients present in the diet may ultimately be wasted if the efficiency of any of these three basic processes is reduced.

Some of the nutrients in the diet are provided in 'chemical' form, e.g. the amino acid lysine (Table 3.2). Since lysine is particularly important to pigs, it is produced industrially (both by chemical synthesis and by fermentation). Manufactured lysine is *identical* to that found as a component of dietary protein and, of course, the pig makes no distinction whatsoever. Some of the vitamins provided in the supplement (last item in Table 3.2) will also have been synthesized, and many will have been at least purified and stabilized. Such nutrients are utilized with far greater efficiency than the same nutrients present as components of 'natural' ingredients. Many of the objections raised by consumers to the use of 'chemicals' in animal feeds therefore seem rather irrational. In fact, manufactured lysine is a common component of the 'slimming' products consumed by many of us, and vitamin concentrates containing exactly the same substances found in pig supplements can be purchased from high-street chemists.

Biologically, these compounds would be unable to substitute for their naturally occurring counterparts if they were not chemically identical to them.

An additional recent debate centres on the use of *particular* ingredients in animal diets. Logically, however, dietary ingredients should be regarded simply as the means whereby animals are fed nutrients, the origin of the ingredients being immaterial. Statements about the pig being 'vegetarian' (herbivorous) are quite wrong—if allowed access to fields, they have a particular liking for earthworms.

3.4.1 Disease

Because they are kept inside to protect them from extremes of climate, many farm animals are confined in close quarters. Such animals are at considerably greater risk from disease than they would be if kept in isolation or under conditions of lower stocking density (humans are no different in this respect). A disease condition which is **subclinical** (i.e. showing no obvious symptoms even to an expert, in this case a veterinary surgeon) may nevertheless cause a reduction in *performance* (e.g. growth) of considerable economic significance. This raises the problem of controlling pathogenic organisms (bacteria, viruses, etc.), including those which may not be regarded as serious.

Several means of control are available, all of which must be considered together. These include providing the animals with food which is as free as possible from pathogens, purchasing disease-free stock in the first place and keeping stock as isolated as possible from other animals. By other animals we mean animals of the same species, domestic pets, wild mammals and birds (which are important reservoirs of infectious disease) and humans (other than those working on the farm)—since the average human is surprisingly unhealthy.

The diagnosis of livestock diseases and their cure have assumed fundamental importance in recent years. This is not because animals are less healthy than they were, but because the economic consequences of unhealthy stock have become more serious as profit margins have been reduced. These days, pigs are reared in comparatively germ-free conditions; being healthier, they tend to perform better. However, they do not really get the chance to build up natural immunity to disease (which requires them to catch the disease first). This means that they will be *more* susceptible to whatever illnesses are around (hence the need to keep them in comparatively germ-free conditions!).

In these circumstances, therapeutic agents are obviously of value and, in all likelihood, some will be contained in the dietary supplement (Table 3.2). However, their use is strictly controlled. Only those *not* used in human medicine may be administered to animals without a veterinary prescription, and then only at low levels. In particular, broad-spectrum (i.e. relatively non-specific) antibiotics are forbidden for routine use in pig diets.

Question 3.3 Re-read the last two sentences. What is the biological reasoning behind these restrictions? (Think in terms of natural selection and resistance.)

3.5 Qualitative aspects of pig-meat

We have already begun to think about the issue of product 'quality' in connection with egg yolk colour (Section 3.2).

Question 3.4 What qualitative aspects of pig-meat do you think might be of importance to consumers, retailers and ultimately producers?

As societies become increasingly affluent, and hence able to view food more in terms of pleasure than of sustenance, issues of quality acquire greater significance. Moreover, as we have seen with eggs, quality may mean different things to different people and it is also probably true to say that perceptions of quality are changing ever more rapidly. To what extent can producers (i.e. farmers) respond to these increasing and changing demands? Certainly quality *can* be influenced at the point of production, but a range of production systems sufficient to cater for all the various and sometimes conflicting quality demands could not possibly be provided within one farm. This point is accepted to some extent by the retail sector, which expends considerably more effort varying products by processing them than it does trying to influence their production. In any case, is it possible to modify the quality of a product *generally*—or is there a risk that by improving one aspect, another may suffer? Indeed, the *assessment* of quality may be quite difficult. While it may be fairly easy to measure the fat content of meat (if only by visual assessment), this is certainly not true of its organoleptic characteristics.

3.5.1 Pig-meat quality

In the past, pig-meat quality has almost invariably been viewed in terms of its fat content. In the early years of the 19th century William Cobbett wrote in *Cottage Economy* that:

> *Lean bacon is the most wasteful thing that a family can use. In short it is uneatable except by drunkards who want something to stimulate their sickly appetite.*

Comments like this contributed to the interest in fat production. Fat has a very high calorific value and it was undoubtedly an important source of energy for the harsh physical work demanded of the labouring population. In addition, the fat content of meat improves its succulence and aids its preservation during curing (e.g. in the case of bacon). The pig accumulates fat comparatively easily and this was regarded as particularly important in an industrial context. At the end of the last century, for example, the Great Plains of the USA supplied the food which pigs converted into fat and which was then dispatched to Europe for a variety of uses. Whole towns sprang up which were dependent on this trade, chief of which was probably Porkopolis (now known, rather more politely, as Cincinnati). However, the emphasis on fat for human consumption was not ubiquitous and it may not even have been regarded as all that important. A few years before Cobbett, John Lawrence wrote:

> *That fat is the labourers' meat is of that species of argument generally pressed into the service of a favourite hypothesis; from the widest enquiries I made, they invariably rejected over-fatted mutton, where they have any choice, for which there is this good reason; they must eat up all they buy whereas the rich may admire, taste and throw the remainder away. The late fashionable excess in fattening even pork and bacon has generally disgusted the labourers.*

This interesting social commentary is as valid now as it was then—considerations of meat quality become considerably more important with increasing affluence. It also serves to underline that definitions of quality were, and still are, variable.

In fact, the current emphasis on lean meat was paralleled by a trend away from fat even in the last century. Despite the obsession livestock producers appeared to have with excessive fat cover, the general population was not convinced. Evidence for this includes the following comments made by pig judges at Royal Shows, as recorded in the *Journal of the Royal Agricultural Society*:

> *... all blubber with little of the lean flesh that is wholesome for man.* (1882)

> *... with the sole exception of the ability to fatten, nearly all the points of excellence have been improved out. As producers of lard they are unsurpassed but from the bacon curers' point of view they are fast becoming impractical.* (1889)

Today's emphasis on a reduction in the fat content of meat is, to a large extent, a reflection of the perceived link between fat consumption and disease in the human population.

3.5.2 Measurements of pig-meat quality

While visual appraisal may be adequate for consumers, assessment of meat quality within the industry has to be quantified for two main reasons. The first is that, if production systems are to be modified to improve a particular quality characteristic, then its biochemical and physiological bases have to be identified; an essential step in this identification is usually quantification of the characteristic. The second is that, since improved quality is invariably associated with increased price to the consumer, it is necessary to be able to demonstrate objectively that the quality of one sample is higher than that of another.

In the UK, the content of fat in pig-meat is usually evaluated in terms of the depth of the back-fat (i.e. subcutaneous fat) and the amount of lean is usually determined by measuring the size of the 'eye muscle' (Figure 3.2).

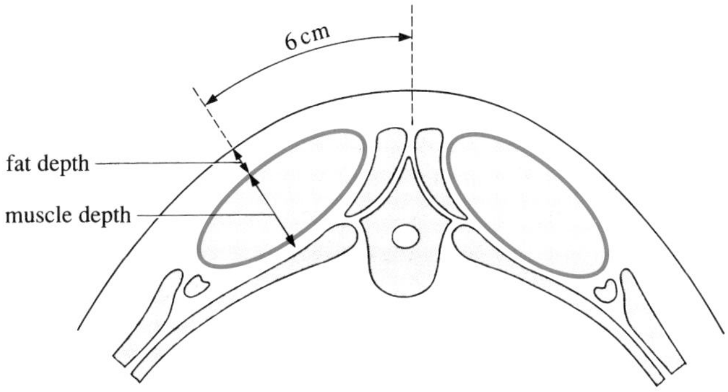

Figure 3.2 Transverse section through the lumbar region of a pig showing where the key measurements of meat quality are taken between the 3rd and 4th from last ribs.

Organoleptic characteristics such as 'juiciness', 'tenderness' and 'flavour' are assessed by taste panels using linear scales ranging from, for instance, (1) 'dry' to (8) 'succulent/juicy'. Despite the use of such scales, these appraisals are still fundamentally subjective and thus quite difficult to analyse.

Interestingly, some of these organoleptic features are positively related to the fat content of lean meat. For instance, in a recent test conducted by the Meat and Livestock Commission, increased fat content of muscle (from 7.8 to 8.2 g fat per kg muscle) was found to be related to improved 'tenderness' (from 4.86 to 5.07 on its linear scale).

The fat associated with muscle can lie *between* discrete muscles (*inter*muscular fat) or *within* individual muscles (*intra*muscular fat). Both contribute to overall muscle fat. Although it may seem difficult to assess muscle fat content without recourse to chemical techniques, visual appraisal is *qualitatively* quite accurate.

▷ Which of the bacon slices shown in Plate 3.2 has the more intramuscular fat?

▶ The lower one.

The striations of fat within muscles are known colloquially as 'marbling'. Generally speaking, the more extensively marbled it is, the 'juicier' the meat (and the easier it is to cook!). However, this is an example in which the presence of a particular component (i.e. fat) can have both positive (taste) and negative (high fat intake) quality aspects. How can the producer possibly reconcile them?

Animal fats and plant oils are chemically rather similar to one another, the major difference being that animal fats are more 'saturated' and plant oils more 'unsaturated'.

Question 3.5 What do the terms 'saturated' and 'unsaturated' mean in this context? (Refer back to Chapter 2 if you need to.)

Although these differences can be chemically quantified, visual appraisal is again useful (at least in a qualitative sense) since generally fats are solid at room temperatures while oils are liquid.

The inclusion of a greater proportion of polyunsaturated fats in the human diet has been widely recommended. To a comparatively minor extent, it might be possible to accommodate this change by altering the chemical composition of the fat within meat. However, the greater the degree of unsaturation the greater the chemical instability of the fat. This instability shortens the meat's shelf-life because off-odours are increasingly produced as the fat deteriorates. Relatively unsaturated fat is also softer, which consumers appear to dislike since the meat seems 'greasier'. Once again, the producer is faced with a dilemma. How can a commodity be produced which is perceived as 'healthier' without it becoming unacceptable to the consumer from an organoleptic point of view?

3.5.3 Modifying pig-meat quality

Fat content, its chemical structure and its distribution can thus have a marked influence on overall meat quality. Producers of pig-meat have two major means at their disposal to alter, both qualitatively and quantitatively, the fat content of pig-meat: *breeding policy* and *nutrition*.

Breeding policy

In any population, there is variability in form. Some characteristics, such as coat colour, are determined by relatively few so-called 'major' genes; individuals will tend to have one or another of relatively few possible colours or colour combinations. Other characteristics, such as fat content, are determined by several or many 'minor' genes acting together; individuals are therefore likely to range from being very lean to very fat. The degree to which a characteristic such as leanness is passed on from parent to offspring is referred to as the **heritability** of that characteristic.

Having identified a significant feature which might be modified by breeding (e.g. fat content), the first stage of an animal breeding programme is to estimate the **variability** of the feature within the population. For many features of interest, variability can be represented as a familiar 'bell'-shaped **normal distribution** curve, in which the bulk of measurements are clustered around the **mean** ('average') value with more extreme measurements tailing off in both directions (such a distribution was encountered in the answer to part (a) of Activity 3.1). The 'flatter' the curve, the more variable the feature under consideration. The extent of this variability can be summarized numerically as the **standard deviation** (or **variance**, which is the standard deviation squared) of that feature in the population.

The next stage in a commercial breeding programme would be to breed only from those animals which are above average for the desired characteristic (in this case, leanness). The general principles are illustrated in Figure 3.3.

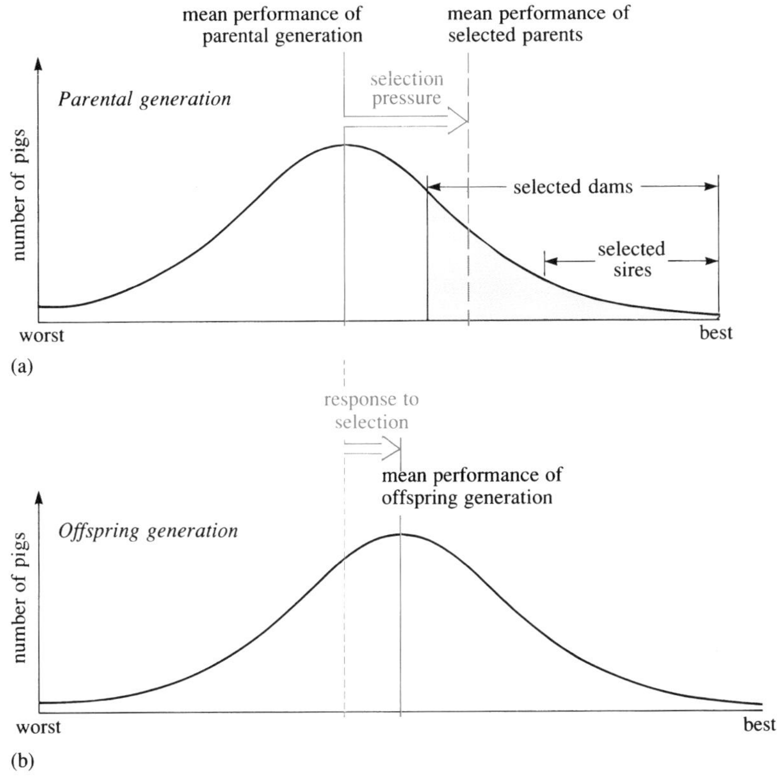

Figure 3.3 General principles of animal breeding. (a) The distribution of the characteristic in question (e.g. leanness) in the parental generation; also shown are the distributions of the characteristic in the selected dams (female parents) and selected sires (male parents), the mean of the characteristic in the selected parents (i.e. their mean performance) and the selection pressure applied (i.e. the difference between the mean performances of the selected parents and of the parental generation as a whole). (b) The distribution of the characteristic in question in the offspring generation; also shown is the response to selection (i.e. the difference between the mean performances of the offspring and parental generations).

Question 3.6 According to Figure 3.3, in what respects do selected dams and selected sires differ for the characteristic being selected? How would you explain these differences?

Because each selected sire can fertilize several selected dams, much greater **selection pressure** can be exerted on the sires, i.e. a smaller number of 'better', less variable animals can be used. If the same selection pressure were applied to the dams, many fewer young pigs would be produced.

Three factors influence the response to selection, i.e. the extent to which the population mean for the character in question changes between the parental and offspring generations. These are:

1 The *heritability* of the particular characteristic, i.e. the extent to which the characteristic is inherited by the offspring.

2 The *selection pressure* applied, i.e. the difference between the mean of the population and the mean of the selected parents.

3 The *variability* of the characteristic in the population as measured by its standard deviation or variance.

The greater the heritability, the selection pressure applied and the variability present in the population, the greater is the likely response to selection.

Before a large-scale commercial breeding programme is embarked upon, it is necessary to estimate the heritability of the feature experimentally to ensure that the programme has a reasonable chance of success. Even though the ultimate objective may be to produce leaner pigs, it is usual in these experiments to select two lines over several generations — in this case, a low-fat line and a high-fat line — and also to maintain a control line. The results of such an experiment with Duroc pigs are shown in Figure 3.4.

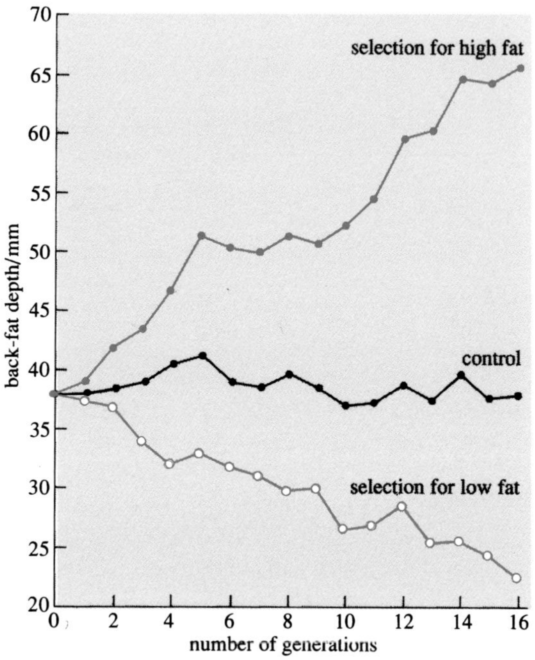

Figure 3.4 Changes in back-fat depth of Duroc pigs over successive generations of selection.

Question 3.7 What is the purpose of the control line? What advantage might there be in having data from both a low-fat line and a high-fat line?

Figures from the Meat and Livestock Commission show that depth of back-fat in pigs between 60–80 kg carcass weight has declined from around 18 mm in the mid-1970s to a current value of below 12 mm.

▷ Can animal breeding programmes be given the entire credit for this change?

▶ Not necessarily. In fact, nutritional changes have also made an important contribution.

The breeding programme to reduce back-fat was both long and expensive. Could subcutaneous fat be altered less expensively? What about simply trimming off the surplus during preparation of the meat? This would have the further advantage that the fat would be available for other purposes. Unfortunately, this option is not without its own problems. It seems that many consumers prefer to purchase pig-meat with the rind (skin) still on. This preference is apparently based solely on visual appeal (for the rind has no nutritional, and very little organoleptic significance, at least in bacon slices and pork chops where most consumers would discard the rind). Yet it is probably the main reason why a simpler and much cheaper solution to the problem of back-fat in pigs could not be found. Ironically, it now seems that low back-fat depths are associated with a tendency for the back-fat to separate from the muscle below it — which also makes the meat visually unattractive. Pigs have been made too lean!

As animals grow, they deposit different tissues in different proportions. When animals are younger, gain in weight is attributable in the main to increases in bone and muscle. Fat deposition predominates as the animals get older (in particular, after puberty). Fat is thus regarded as a **late-maturing tissue**.

▷ How else, then, can selection for leanness be regarded?

▶ As selection for later maturity. Late-maturing animals are those that deposit fat at higher body weights than early-maturing animals. The breeding objective can therefore be regarded as being to delay the onset of rapid fat deposition for as long as possible.

It would, however, be incorrect to regard *all* fat tissue as the same in this respect. The difference between back-fat and marbling fat has already been outlined. It seems that marbling fat is even later maturing than back-fat.

▷ What are the consequences of this for the pig breeder?

▶ Attempts to reduce back-fat are likely to reduce marbling fat even more. As mentioned above, this reduces the 'juiciness' of the meat and hence has a detrimental effect on eating quality.

In these circumstances, it is clear that further progress towards fat reduction in existing breeds is inappropriate. A possible solution to the problem of increasing marbling (a comparatively late-maturing tissue) relative to back-fat (a comparatively early-maturing tissue) would be to identify those breeds that have high amounts of marbling fat and to breed this characteristic into other breeds. This approach is currently of much interest in pig breeding.

Of course, genetic variation in features such as the amount and distribution of fat must result from discrete biochemical differences between animals. Studies of animal growth at the molecular level are therefore assuming considerable importance in the hope that greater understanding of these fundamental processes will allow greater control of growth, with the ultimate aim of improving product quality.

Nutrition

Excessive *amounts* of fat are unlikely to be deposited in pigs provided (a) they are not fed large quantities of high-energy food, (b) they are from comparatively late-maturing breeds and (c) they are slaughtered before reaching puberty (i.e. before much fat deposition occurs). Although most fats in the diet are fully digested and new fats re-synthesized, this is not true of all fats. Some avoid chemical breakdown during digestion and may be deposited directly as carcass fat. Thus, the *chemical composition* of carcass fat may be influenced by diet. In general, the more unsaturated the dietary fat, the greater the degree of unsaturation of the carcass fat.

The effects of diet on carcass quality have been known for some time. One of the earlier attempts to quantify the relationship was with pigs fed on waste peanuts in the southern USA. Although the pigs grew well, their carcass fat was excessively soft. In a more recent experiment in the UK, two groups of pigs were treated similarly except that 'fat A' was added to the diet of one group and 'fat B' to the diet of the other group. The fatty acid profiles of the two fats were quite different as shown in the pair of columns to the left in Table 3.4.

Table 3.4 Influence of degree of saturation of dietary fat on chemical composition of pig carcass back-fat.

Fatty acids*	Fatty acid profile of fat added to diet/g fatty acid kg^{-1} fat		Fatty acid profile of carcass back-fat/g fatty acid kg^{-1} fat	
	Fat A	Fat B	Fat A-fed pigs	Fat B-fed pigs
Saturated				
C12:0	<10	58	0.8	14.3
C14:0	<10	83	15.3	50.3
C16:0	126	358	249.7	296.4
C18:0	42	128	152.4	160.1
Unsaturated				
C16:1	<10	<10	17.4	32.0
C18:1	255	305	336.1	357.2
C18:2	519	67	183.2	65.2
C18:3	57	<10	24.2	13.2
unsat. : sat. fatty acid ratio	4.5	0.6	1.3	0.8

* Each fatty acid is given in the short-hand form (introduced in Section 2.2).

Question 3.8 Which of fat A and fat B is the more unsaturated?

The fatty acid profiles of the carcass back-fat of the two groups of pigs were subsequently determined by chemical analysis (see the right-hand pair of columns in Table 3.4).

Question 3.9 The addition of which of the two fats to the pigs' diet resulted in the more unsaturated carcass back-fat?

The higher ratios of unsaturated to saturated fatty acid concentrations in the diet and in the back-fat were positively correlated with softness of the carcass fat. However, in addition to this overall softness, the effect of one specific fatty acid (linoleic acid, C18:2) on carcass quality is of particular significance. This polyunsaturated fatty acid (which is 'essential' in the diets of pigs as well as humans) is comparatively unstable and, at high levels in meat, may be associated with a reduction in its organoleptic properties. While there is no single point at which problems can be said to occur, it is generally accepted that levels in excess of $150\,g\,kg^{-1}$ back-fat are not to be recommended. The linoleic acid level in the pigs fed fat A (Table 3.4) would therefore be regarded as unacceptable. Although a switch from saturated to polyunsaturated fatty acids in the human diet has been suggested, it would seem that attempting to change the composition of meat fat is perhaps not the best route by which this might be achieved.

3.5.4 Conclusions on pig-meat quality

Consumption of animal fat has allegedly been identified as a causative factor in the incidence of a variety of human diseases. The pig industry has responded by reducing the fat content of pig-meat through both genetic and nutritional means. Definitions of meat quality are currently moving more towards organoleptic and visual appraisal of the product. Although of limited nutritional significance, these characteristics are extremely important in influencing consumer purchasing habits.

Although meat producers are attempting to respond to these developments, they have met with comparatively little success because it is difficult to improve the flavour and taste of meat significantly during production. What happens to meat after slaughter is, however, quite another matter. Variables ranging from post-slaughter storage right through to processing as part of ready-to-cook meals are much more important. A simple proof of this would be to cook and eat a small piece of lean pork in isolation from anything else — no accompanying food, no herbs, no apple sauce, no gravy, no anything. See what it tastes like!

Summary of Chapter 3

1 Animal production may be viewed in terms of optimizing those biological processes responsible for the products or outputs required, e.g. ovulation (providing eggs) and growth (supplying meat).

2 Inputs to animal production systems include the animals themselves, their nutrition, their environment and their health.

3 Egg yolk colour is an example of a quality character that has little significance for nutrition or taste; its visual appeal has become linked with its perceived degree of 'wholesomeness'.

4 Egg yolk colour is determined by pigmenting agents in the chickens' diet; these may be natural, 'nature-identical' synthetic compounds or 'nature-related' synthetic compounds.

5 Egg-laying is often manipulated by artificially controlling the photoperiod to which chickens are exposed.

6 Optimum diets have been determined for pigs in terms of both nutrients and energy-yielding ingredients; some of the nutrients (e.g. lysine) are produced industrially, but are identical to their naturally occurring equivalents.

7 Pigs are kept in comparatively germ-free conditions, as a result of which they have little natural immunity to disease; therapeutic agents are therefore included in their diets. Care has to be taken that microbes do not build up resistance to antibiotics, particularly those used in human medicine.

8 Fat content is an important aspect of pig-meat quality. The amount of fat has been reduced considerably compared to a century or more ago. There is also now a greater emphasis on unsaturated fats in the human diet. These changes illustrate the often conflicting nature of quality considerations: lean meat is also less 'juicy'; a greater degree of unsaturation shortens shelf-life.

9 The fat of pig-meat is amenable to change as a result of breeding policy (selecting both for leanness and late maturation) and nutrition (avoiding large quantities of high-energy food and increasing the proportion of unsaturated fatty acids in the diet).

10 As a consequence of increasing affluence, product quality is assuming greater importance. Quality may have nutritional relevance, it may be important organoleptically or it may simply relate to visual appeal. Quality itself is often poorly defined and, in any case, it is often difficult to fulfil all quality objectives simultaneously.

4 The 'nitrate problem' (4hrs)

4.1 'I only want my rights!'

Everyone needs adequate food *and* clean water. These can properly be seen as basic human rights, even if two-thirds of the world's population does not enjoy them. We in the UK, however, have grown used to a *plentiful* supply of *cheap* food and *unlimited* water that is good enough for drinking and cooking. We now see *these* as rights. But these 'rights' have unfortunately begun to conflict with one another. The root of the conflict lies in the 'nitrate problem'. Nitrate (NO_3^-) is a very ordinary, naturally occurring chemical substance which is an important plant nutrient, but which is also a nuisance and possibly a hazard in the wrong place at the wrong time. The concentration of nitrate in water in many rivers, lakes and aquifers (i.e. natural underground sources of water) has been increasing at the same time as the amount of nitrogen fertilizer used by farmers has increased (compare Figures 4.1 and 4.2). This coincidence may suggest a link between the two, but have a go at Activity 4.1 before you jump to a conclusion.

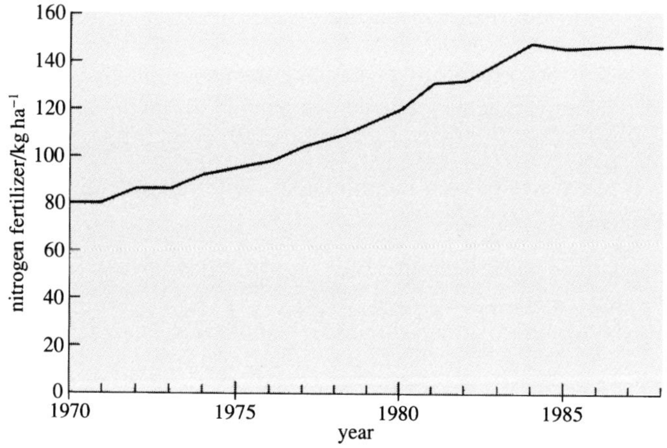

Figure 4.1 The average amount of nitrogen fertilizer applied per hectare per year to all crops and grass, for 1970–1988.

Activity 4.1

Three pairs of *true* statements are given below. For each pair, suggest how the two statements *might* be related to one another, either directly or indirectly.

(a) *Statement 1* The population of storks in Europe has been declining for several decades.

Statement 2 The human birth-rate in Europe has also been declining for several decades.

(b) *Statement 1* Around the turn of the century there was an increase in the number of clergy in England and Wales.

Statement 2 Also around the turn of the century there was an increase in the sales of whisky in England and Wales.

(c) *Statement 1* There has been a large increase in the use of nitrogen fertilizer during the last 30 years.

Statement 2 Nitrate concentrations in natural waters have also increased during the last 30 years.

This chapter aims to show what the 'nitrate problem' is, how the activities of farmers *and other people* have contributed to it and how scientists have tried to find its real causes so that it can be controlled.

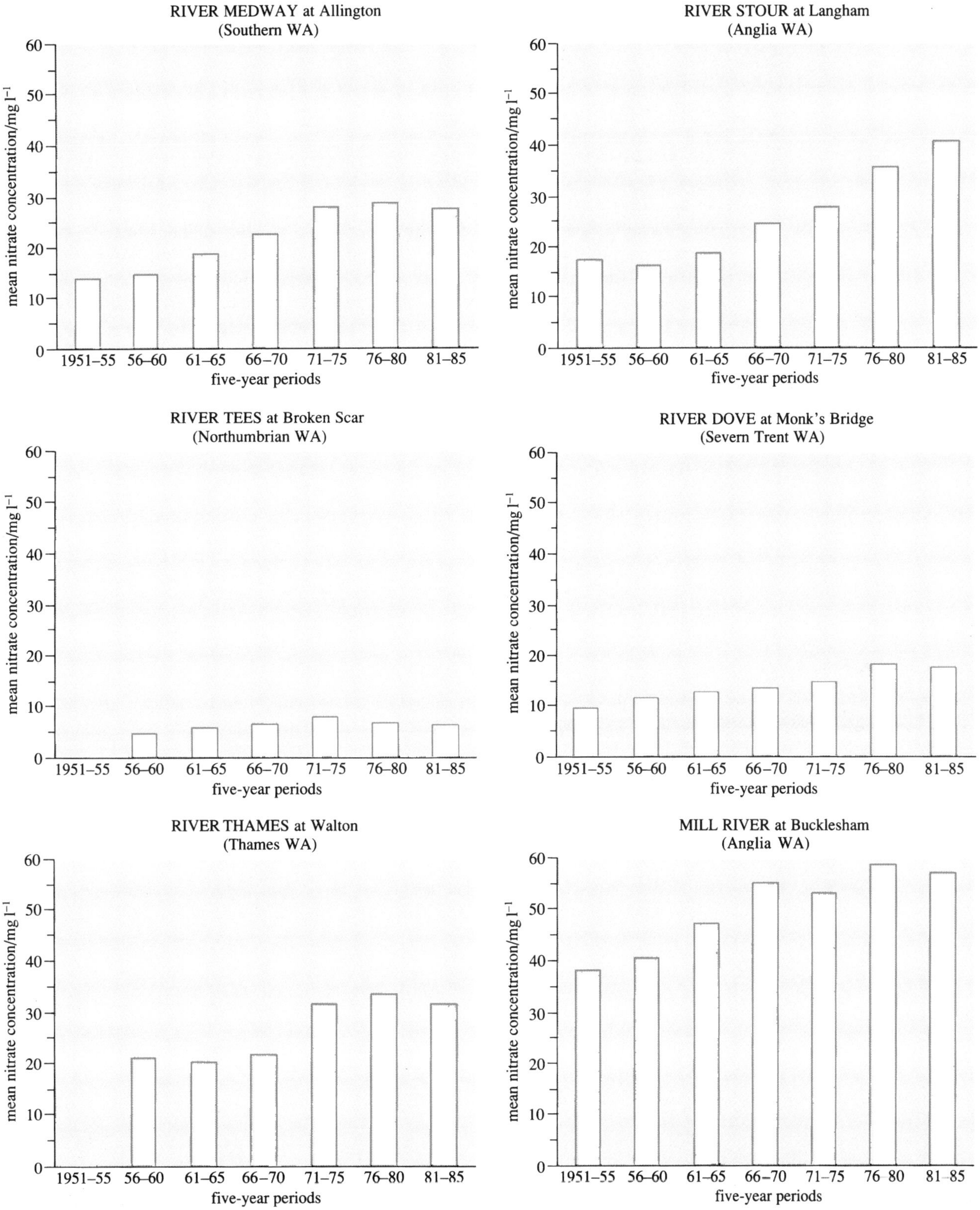

Figure 4.2 Five-year mean nitrate concentrations in six selected English rivers, for 1951–1985. (WA = Water Authority.)

4.2 Why worry about nitrate in water?

People certainly worry about nitrate in water, but do they need to? What problems can it cause, and how much of a threat are these problems? In fact, associated with nitrate in water are two possible health problems (Sections 4.2.1 and 4.2.2) and a general environmental problem (Section 4.2.3).

4.2.1 Methaemoglobinaemia

The first possible health problem is methaemoglobinaemia, also known as 'blue-baby syndrome'. This occurs when infants less than about one year old ingest too much nitrate. The nitrate is converted to nitrite (NO_2^-) in their stomachs and this gets into the bloodstream, where it interferes with the ability of the haemoglobin in the blood to carry oxygen round the body. For the chemically minded, normal oxyhaemoglobin, which contains iron(III) (Fe^{3+}), becomes methaemoglobin, containing iron(II) (Fe^{2+}). The result is that the unfortunate infant is, in effect, suffocated—hence the name 'blue-baby syndrome'. There seem to be two possible reasons why very young children are affected. The first is that their stomachs are not acidic enough to inhibit the bacterial conversion of nitrate to nitrite; when they are older the acidity is too great for this conversion to happen. The second is that foetal haemoglobin, which persists in the blood up to the age of about one year, has a greater affinity for nitrite than has normal haemoglobin.

Activity 4.2

Carefully read through the following information which relates to methaemoglobinaemia.

- Of two cases that occurred in 1950, *The Lancet* reported that 'there were diarrhoea and vomiting and the child's complexion was slate-blue' (case 1) and that 'blood drawn from a vein was a deep chocolate-brown' (case 2).
- Both cases occurred when the baby's feed had been prepared using water from a well. In the USA the condition is usually known as 'well-water methaemoglobinaemia'. In a large proportion of the known cases, babies were being given water from privately dug wells.
- The last case in Britain was in 1972 and the last fatal case (case 1 above) in 1950.
- In case 2, the water was 'heavily contaminated with coliform organisms' (*E. coli*). These organisms cause gastroenteritis, which greatly exacerbates the effects of the nitrate in water.
- Nearly all cases have resulted from nitrate concentrations greater than $100\,\mathrm{mg\,l^{-1}}$. Case 1 resulted from a concentration of $200\,\mathrm{mg\,l^{-1}}$. Although the concentration in case 2 was only $95\,\mathrm{mg\,l^{-1}}$, the water was contaminated with *E. coli* (see above). Much greater concentrations have occasionally been involved, e.g. $1\,200\,\mathrm{mg\,l^{-1}}$ in one case in the USA.
- The European Commission's limit for nitrate in domestic water supplies is $50\,\mathrm{mg\,l^{-1}}$.

On the basis of the above information, decide whether *you* consider the risks of methaemoglobinaemia to be serious. Before reading the answer, which does no more than give another person's view, *write down* your decision and the reasons for it.

4.2.2 Stomach cancer

Among the cancers, that of the stomach is the second largest killer of both men and women; only lung cancer kills more men and only breast cancer kills more women. Stomach cancer is a painful and debilitating way to die.

It has been suggested that nitrate in water causes stomach cancer when it is converted to nitrite. In the stomach, the nitrite reacts with a chemical compound known as a secondary amine (which could come, for example, from the breakdown of meat). The product of such a reaction would be an *N*-nitrosamine, a type of compound known to cause cancer.

▷ What is the weak link in this suggested chain of reactions?

▶ The conversion of nitrate to nitrite. There would not be enough time for this to happen in the mouth and it would not be expected to happen in an adult's stomach because of the level of acidity.

Activity 4.3 *You should spend up to 30 minutes on this activity.*

Consider the information given below and the data in Tables 4.1 and 4.2. Then decide *for yourself* whether there is any real risk that nitrate in water could lead to stomach cancer. It is important to put *your own thoughts* down on paper (if only in note form) *before* reading the answer (which gives the personal view of someone who has given a lot of thought both to the nitrate problem and to cancer).

Much of the information consists of tabulated data, about which you should think critically. Take particular note of the results of the statistical tests performed on the data. These are given as the probabilities of differences at least as large as those reported *occurring by chance*. Conventionally, the difference occurring by chance is regarded as a distinct possibility if the probability (P) is 1 in 20 ($P = 0.05$) or greater ($P > 0.05$); in these circumstances the difference is said to be *statistically non-significant*. A probability of less than 1 in 20 ($P < 0.05$) means that the difference is less likely to have occurred by chance and such a difference is said to be *statistically significant*. Successively lower probabilities—such as less than 1 in 100 ($P < 0.01$), less than 1 in 1000 ($P < 0.001$) and less than 1 in 10000 ($P < 0.0001$)—are associated with successively lower likelihoods of the difference occurring by chance and hence with successively greater statistical significance.

It is also important to think critically about how the data were collected. Could they contain any form of bias? Could some important factors have been neglected?

- During the last 30 years nitrate concentrations in water supplies have increased generally, but the incidence of stomach cancer has declined.
- In the early 1980s a group of **epidemiologists** (i.e. researchers on the occurrence and cause of disease in populations) at the Radcliffe Infirmary in Oxford mounted a detailed study of the relationship between nitrate and stomach cancer. They selected two areas (I and II) in which the incidence of stomach cancer was unusually low and two areas (III and IV) in which it was unusually high. They arranged for samples of saliva to be collected from healthy *visitors* to the hospitals in each of these areas (i.e. *not* from patients). These samples were analysed and a summary of the results is given in Table 4.1. The 'refined sample' refers to people who had not eaten or drunk anything for two hours before giving the sample, a precaution which gives a better indication of the background level of nitrate in their bodies.

- If exposure to nitrate and stomach cancer are linked, one group of people particularly at risk should be those who work in factories producing ammonium nitrate (NH_4NO_3) fertilizer. Although modern fertilizer plants are clean and equipped with dust extractors, you always get a salty taste on your tongue from traces of ammonium nitrate in the air. Those working in these plants must absorb more nitrate than other factory workers. A long-term epidemiological study was made in which the health risks to workers in an ammonium nitrate fertilizer plant were compared with those to workers in other jobs. The main statistic used was the mortality (i.e. number of deaths) during the period of the study. Distinction was made between 'heavily exposed' and 'less heavily exposed' fertilizer workers (a survey of current workers at the plant having established that nitrate and nitrite concentrations in saliva increased with degree of exposure to nitrate). In each case, the mortality was compared with the mortality that would have been expected on the basis of the number of deaths among comparable workers in the locality (Table 4.2).

Note that both of the possible health problems associated with nitrate in water, methaemoglobinaemia and stomach cancer, arise from the conversion of nitrate to nitrite. *Nitrate itself is not a threat to health.* However, in some circumstances it can pose an environmental threat.

Table 4.1 Mean nitrate concentrations ($\mu mol\,l^{-1}$) in saliva samples from people living in areas of low (I and II) and of high (III and IV) incidence of stomach cancer. The 'refined samples' were from people who had not eaten or drunk for 2 hours before giving the sample. (n.s. = difference between the means is statistically non-significant, i.e. $P>0.05$.) (Source: adapted from Forman, D. *et al.*, 1985, 'Nitrate, nitrate and gastric cancer in Great Britain', *Nature*, **313**, pp. 620–625.)

	Low incidence areas			High incidence areas			Areas combined		
	I	II	Statistical significance of difference between I and II	III	IV	Statistical significance of difference between III and IV	Low incidence areas (I and II) combined	High incidence areas (III and IV) combined	Statistical significance of difference
all samples	208.3	157.3	$P<0.01$	107.2	107.9	n.s.	—	—	—
refined samples	171.5	149.5	n.s.	97.0	116.8	n.s.	162.1	106.3	$P<0.0001$

Table 4.2 Mortality from cancer of the stomach and other causes among 1 327 male workers in an ammonium nitrate fertilizer plant from 1 January 1946 to 28 February 1981. The numbers of deaths *observed* among those heavily and less heavily exposed to nitrate are compared with the numbers of deaths that would have been *expected* from local population statistics. (n.s. = difference between the observed and expected numbers is statistically non-significant, i.e. $P>0.05$.) (Source: adapted from Al-Dabbagh, S. *et al.*, 1986, 'Mortality of nitrate fertilizer workers', *British Journal of Industrial Medicine*, **43**, pp. 507–515.)

Disease	Heavily exposed workers			Less heavily exposed workers			Both groups of workers together		
	Observed	Expected	Statistical significance of difference	Observed	Expected	Statistical significance of difference	Observed	Expected	Statistical significance of difference
stomach cancer	7	7.22	n.s.	5	4.84	n.s.	12	12.06	n.s.
all cancers	59	51.36	n.s.	32	35.47	n.s.	91	86.83	n.s.
respiratory diseases	21	30.97	n.s.	15	20.07	n.s.	36	51.04	$P<0.05$
heart disease	56	67.64	n.s.	36	45.98	n.s.	92	113.72	$P<0.05$
all causes	193	219.78	n.s.	111	148.33	$P<0.01$	304	368.11	$P<0.001$

4.2.3 Eutrophication

The general environmental problem associated with nitrate in water arises from excessive nitrate in surface waters such as rivers and lakes, where it causes water plants to grow excessively and algae to 'bloom'. Underwater plants clog conduits and entangle anything passing through the water, from propellers to fishing tackle; too much reed growth blocks waterways and overloads river banks, sometimes causing them to collapse. Algae (some of which may be toxic) form an unsightly scum on the surface of the water and, when they die, the bacteria that decompose them use up so much oxygen from the water that other organisms (notably fish) die too. This is what is meant by **eutrophication**. However, algae also need phosphate and lack of this may limit their growth in freshwater. Much of the phosphate present in surface waters comes from discharges from sewage works.

4.3 Where does nitrate come from?

This section is concerned with the global **nitrogen cycle**. The global cycle is, of course, made up of individual cycles such as those of the ocean, the atmosphere and the soil. However, even the nitrogen cycle of the soil cannot be regarded as a single entity. For example, there are differences between the cycles in tropical rainforest and savanna grassland, and even more importantly between land covered by natural vegetation and land under human management. We shall be paying particular attention to the nitrogen cycle in British farmland (Figure 4.3).

Figure 4.3 A schematic diagram of part of the nitrogen cycle.

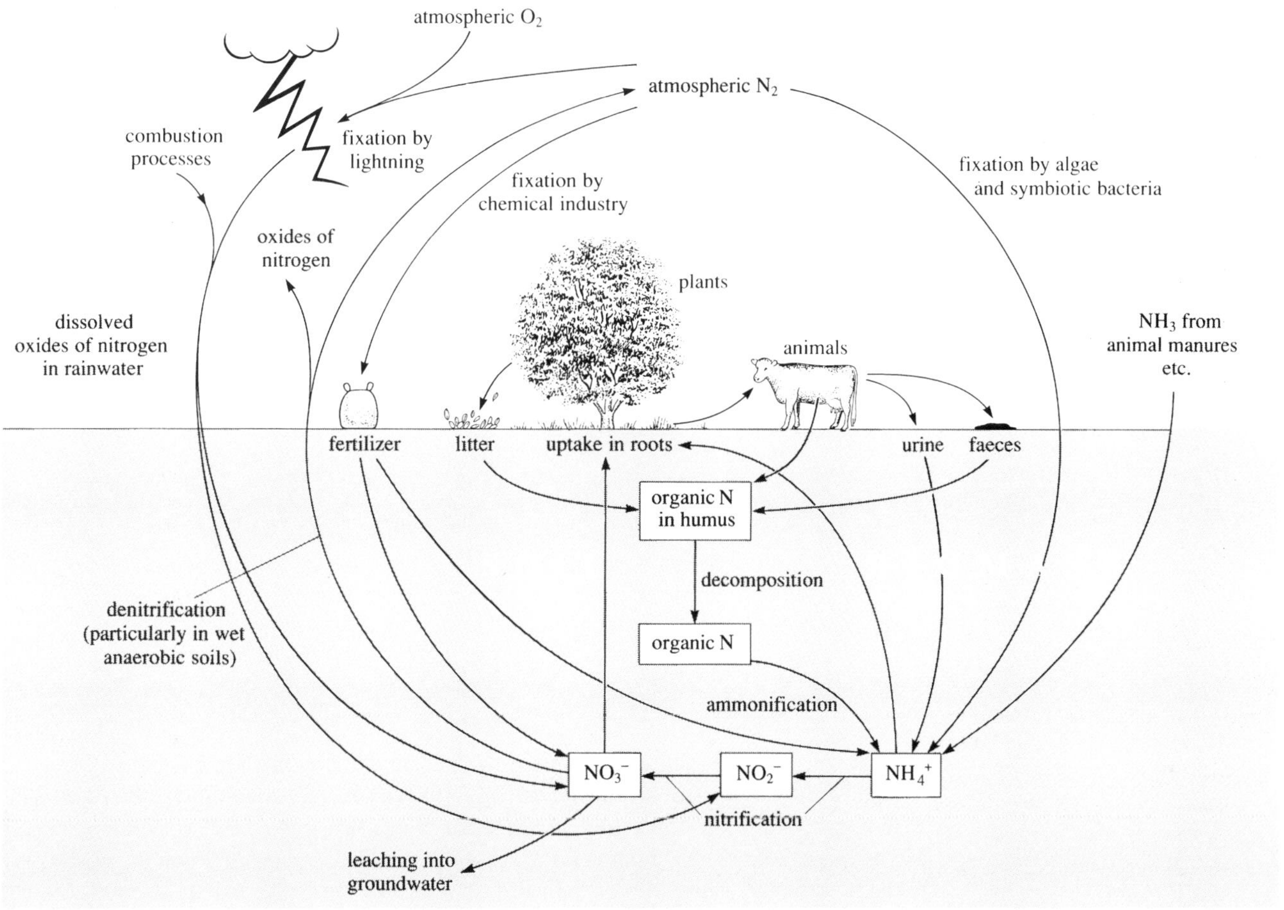

Ultimately, nitrate comes from the capture or **fixation** of nitrogen in the air. Atmospheric nitrogen and oxygen can combine directly in flashes of lightning to form oxides of nitrogen that dissolve in rain to give nitrate and nitrite; in the soil nitrite is oxidized to nitrate. Oxides of nitrogen can also be formed during combustion in petrol engines. However, most nitrate is fixed via ammonium (NH_4^+) by either specialized micro-organisms (microbes) or the chemical industry.

Nitrogen-fixing microbes are of two kinds: (a) certain free-living algae, and (b) bacteria that form symbiotic associations with leguminous plants such as peas, beans and clover. The bacteria live in nodules on the roots of these plants; some of the photosynthetic carbohydrate produced by the plants is utilized by the bacteria, while some of the fixed nitrogen is utilized by the plants. An additional advantage to the bacteria is that the environment of the root nodule allows them to control the concentration of oxygen to which they are exposed, an important consideration because the key enzyme in nitrogen fixation (nitrogenase) is readily deactivated by oxygen.

The chemical industry fixes nitrogen by reacting it with hydrogen to produce ammonia (the Haber–Bosch process); the ammonia is subsequently oxidized to nitric acid and thence converted to nitrate. Taking the world as a whole, microbes currently fix about twice as much nitrogen as the chemical industry, with lightning coming a poor third (Table 4.3). Being a relatively small and quite industrialized country, the UK taken on its own presents a rather different picture—with the chemical industry estimated to fix 1.6 million tonnes per year and microbes only 0.3 million tonnes per year.

Table 4.3 Estimated contributions to nitrogen fixation on a global scale.

Agency	Amount/10^6 tonnes yr^{-1}
microbes	140
chemical industry	74
lightning	8

In fact, nitrogen fixed by lightning may represent a relatively small proportion of the nitrogen reaching the surface of the soil even compared with other inputs from the atmosphere. Excluding the deliberate application of fertilizer, every year each hectare of land in southern and eastern England receives an average of about 35 kg of fixed nitrogen from above—some as ammonium and nitrate dissolved in rainwater and the rest as nitrogen oxides and ammonia deposited directly on to the soil and crop surfaces. The nitrogen oxides and the nitrate may well result from combustion processes well away from the farm (e.g. motor vehicles), but the ammonia is quite likely to be of local origin—especially if there are animals or animal manures on the farm.

A very obvious, though important, point is that (disregarding isotopes for the time being) all nitrate ions are identical and there is therefore no way in which one can tell where a particular nitrate ion has come from. Nitrate found in soil on organically farmed land is thus exactly the same as nitrate found elsewhere.

4.3.1 Nitrate in the soil

'I don't see how you can get interested in dirt', said a friend when I changed from straight chemistry to soil science. The soil, however, is very much more than dirt—it is the basis of life on Earth. One of the main things that makes the soil more than just dirt is its organic matter, both dead and alive. The dead part mainly comprises the remains of plants, some of which lived last year and some tens or hundreds of years ago. Most of these remains are not recognizable as plant material because they have been processed by the soil's living organisms (its **biomass**), and have become **humus**. Humus is what makes most topsoils darker in colour than the corresponding subsoils, and it plays an important part in stabilizing the soil crumbs that are needed for a good seed-bed. It is also a very large repository for nitrogen. On a global scale, only the atmosphere and the oceans contain more nitrogen than does the soil's organic matter (Table 4.4). We shall refer to nitrogen in organic matter (whether living or dead) as **organic nitrogen**. Even a soil that has been cultivated for a long time, and therefore contains relatively little organic matter, will have 2–3 tonnes of nitrogen per hectare in the topsoil. More typical arable soils will have 3–5 tonnes ha^{-1}, and grassland and peat soils will contain much more.

Table 4.4 Storage locations of most of the global nitrogen.

Location	Amount/tonnes
atmosphere	3.9×10^{15}
oceans	2.4×10^{13}
dead organic matter in soil (humus)	1.5×10^{11}
plants	1.5×10^{10}
living organic matter in soil (biomass)	6×10^{9}
land animals	2×10^{8}
people	1×10^{7}

Although present in the soil in vast quantities, the nitrogen in dead organic matter is inert and cannot be **leached** (i.e. washed beyond the reach of plant roots) until it is further processed by soil organisms. It is clear that the biomass of the soil plays a crucial role in the nitrogen cycle.

Think of a herd of cows in a field—tonnes of flesh ambling around. It is important to realize that the mass of living matter under the soil surface is about the same as that of the cows. Instead of perhaps 30 or 40 large organisms you have millions of rather small ones, ranging in size from bacteria and fungi to earthworms. Dead organic matter coming into the soil may initially be attacked by fungi, but more often small soil animals (e.g. earthworms, millipedes and springtails) get to it first. These animals break up the organic matter and mix it more intimately into the soil. They retain some of the nitrogen in their own bodies, add some of it as faeces to the humus and excrete the rest as ammonium. The final stage in this process of **decomposition** is carried out by microbes.

Under the soil's placid surface, nature is indeed 'red in the tooth and claw' as pictured by Tennyson. The predation by some species on others also plays an important part in the overall process of releasing nitrate from organic matter. As in the case of organisms feeding on dead organic matter, the predators absorb some of the nitrogen from the bodies of their prey and return the rest to the soil in their faeces; some of the absorbed nitrogen is also returned to the soil as excreted ammonium, which is usually converted to nitrate quite rapidly.

In most soils ammonium is soon converted to nitrate by specialized bacteria. The production of ammonium from organic nitrogen is known as **ammonification** and its conversion to nitrate via nitrite as **nitrification**. Since ammonium and nitrate are together known as 'mineral N' ($\mathbf{N_{min}}$), these two processes are usually referred to as **mineralization**.

Although the amount of nitrogen applied as fertilizer (possibly 100–200 kg ha^{-1} yr^{-1}) may seem trivial in comparison with the amount in the soil already, it is supplied in a form that is readily available to plants—as nitrate and ammonium. Ammonium, as we have seen, is rapidly converted to nitrate by soil microbes. Nitrate is extremely soluble in water and therefore readily leached. Thus, in the case of soil nitrogen, *availability* to plants is inseparable from *vulnerability* to loss from the soil. Once leached, nitrate becomes a potential problem.

Activity 4.4 *You should spend up to 15 minutes on this activity.*

Before you proceed further, make sure that you know the meaning and significance of the following terms by writing a one sentence definition of each: ammonification; biomass; decomposition; eutrophication; fixation; humus; leaching; mineralization; nitrification; N_{min}. Ideally, you should be able to do this without looking back at the previous text, but do so if necessary.

4.4 Who controls nitrate on the farm?

Would I be right to guess that, until very recently, your answer to this question would have been 'the farmer', and that you would not have thought much further about it? Farmers certainly play their part, but they are very far from being the only ones to add nitrate to the soil. Soil microbes also 'have their say' and farm animals can have a strong influence too. The fact is that nitrate in the soil is controlled jointly by farmers, soil microbes and farm animals.

Farmers add nitrate to the soil in fertilizer. (Fertilizers often contain ammonium too, but microbes usually soon convert it to nitrate.) However, this is not the farmers' only contribution; they also *cultivate* the soil and, in doing so, let in air and increase the vulnerability of the soil's organic matter to decomposition and mineralization by microbes—leading to release of ammonium and thence nitrate.

In most cases the soil has been cultivated regularly for a number of years and the amount of vulnerable organic matter is not particularly high. (In fact, roughly 2–5% of the organic nitrogen is mineralized each year.) Sometimes, however, farmers plough up land that has grown grass, and hence been accumulating organic matter, for several years; in these cases appreciable amounts of nitrate can be released. For instance, large areas of very old grassland were ploughed up during and after the Second World War to increase food production. Recent calculations by members of staff of Rothamsted Experimental Station and the Agricultural Development and Advisory Service (ADAS) suggest that this ploughing contributed appreciably to our current nitrate problem.

Soil microbes also put nitrate into the soil (Section 4.3.1) provided they have access to food, warmth and moisture. The more organic matter that is available to them as food, the more nitrate they produce; hence the significance of ploughing up grassland. The significance of warmth and moisture becomes clear when you think about the soil during the period following harvest in a temperate country such as the UK.

▷ Study Figure 4.4. Does the average daily rate at which water drains through the soil vary much through the year?

▶ Yes. The rate is considerably greater in winter than in summer.

▷ Is this pattern accounted for by more rain falling in winter than in summer?

▶ No. Such variation as there is in rainfall is insufficient to account for the greatly increased rate of drainage in winter.

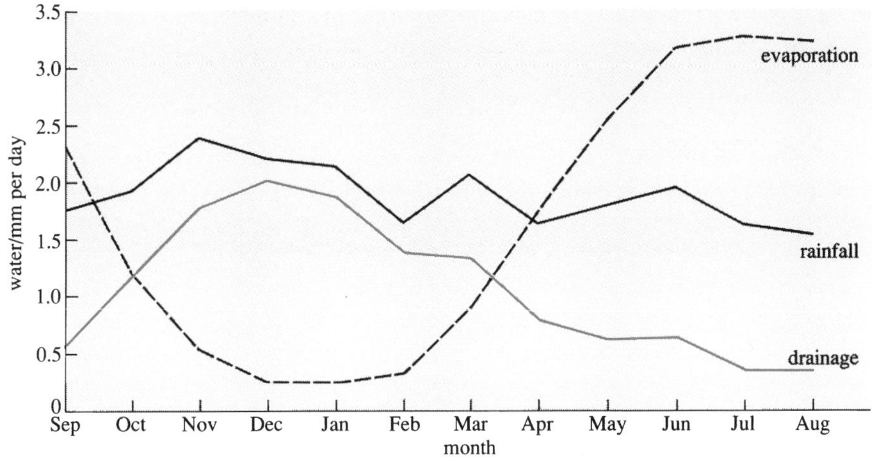

Figure 4.4 Variations during the year in the average daily amounts of rainfall, evaporation and drainage through the soil, in the UK.

▷ What is the main cause of variation in the average daily drainage rate?

▶ The variation in the average daily rates of evaporation of water from the soil. Because of the longer days and higher temperatures of summer, the rates of evaporation are higher—and the rates of drainage lower—than in winter. Shorter winter days and lower temperatures lead to lower rates of evaporation and higher rates of drainage at this time of the year.

In late summer and early autumn the soil is still warm from the summer sun; but, as evaporation fails to keep pace with rainfall, the soil becomes moister, stimulating the microbes to vigorous activity—and hence vigorous production of nitrate. During winter much of the rain drains straight through the increasingly saturated soil, carrying nitrate away with it. This nitrate is leached *because it is there at the wrong time*; either the soil is bare or the crop has only just started to grow and so cannot yet take up much nitrate. This makes the point that we are not really talking about a nitrate problem so much as a problem of **untimely nitrate**. Nitrate in the right place at the right time is essential for our food supply.

Farm animals also contribute to untimely nitrate in the everyday business of living. Like microbes, farm animals contribute to the nitrate problem when they excrete urine. They do not contribute nitrate *directly*; but their urine contains urea, which is rapidly converted to ammonium and thence to nitrate by microbes in the soil. Their faeces too contain various compounds of nitrogen which soil microbes can convert to nitrate. One problem is that a cow wanders around a field, eating here and there, and then urinates on a small patch (usually of about $0.5\,m^2$). The equivalent of up to $1\,000\,kg\,ha^{-1}$ of nitrogen is applied on that patch in a few moments—this is far more than the grass can use *annually*. The cow's faeces also supply much more nitrate than the grass can use. These problems are further exacerbated by the gregarious habits of cows!

Another problem is that cows are not efficient converters of nitrogen. Most of the nitrogen fertilizer applied to grass is taken up by it, but when the grass is eaten by a cow only a small proportion of the nitrogen is converted to organic nitrogen in meat, milk or some other usable commodity; the rest is excreted. Work at the Institute of Grassland and Environmental Research suggests that only about 10% of the nitrogen applied in fertilizer ends up in these commodities; the rest is scattered in concentrated patches around the fields, from which nitrate is likely to be leached. This problem of inefficiency makes the stocking density (i.e. the number of cows per hectare) very important. Within limits, productivity per hectare increases with increasing stocking density (while the productivity per individual cow declines). However, increasing the stocking density also means that more grass, and therefore more nitrogen fertilizer, is needed—and that there will be more patches over-rich in nitrate and hence likely to lose it through leaching.

Can this problem be avoided? Nobody has yet devised a nappy suitable for cows! The only way round the problem is therefore to keep the cows in an area from which the excreta (urine and faeces) can be collected at times when the risk of leaching is high. Keeping them on straw provides the familiar farmyard manure. If they are kept on concrete in a yard, the excreta can be hosed into a tank to give slurry. Both of these products can then be returned to the soil. These approaches certainly have some advantages. For instance, the grass can be fed to the cows either as hay or as silage made from cut grass (a system that uses nitrate very efficiently because so little is leached). The main problems lie in storing and handling the farmyard manure and the slurry—which requires care and some expense—and in returning them to the soil. Both are difficult to spread evenly on the soil and both contain nitrate (and material that soil microbes can readily convert to nitrate), making them vulnerable to heavy rain which could wash them off the soil into the nearest watercourse.

Since leached nitrate represents a wasted resource, the farmer will not contribute to untimely nitrate if this can be avoided. Thirty or forty years ago it was common for fields to be ploughed in autumn and then left bare over winter so that the frost could break up the soil. This practice, together with leafless trees, gave the winter countryside a kind of stark beauty which you can still see reflected in paintings (Plate 4.1). However, it also meant that microbes were stimulated by the ploughing to produce nitrate at a time when there was no crop to remove it, and which was therefore leached. Farmers now grow far more autumn-sown crops, notably winter wheat, and much less bare soil is to be seen in winter than in the past. Reports from at least one water authority have suggested that this change in practice has almost halted the increase in nitrate concentration in water going into aquifers and in some places has reversed the trend. Another helpful change in practice is that some farmers now grow so-called 'catch' or 'cover' crops specifically to mop up nitrate and plough them back into the soil when the risk of leaching is over (examples include rye-grass, oilseed rape, radish and white mustard).

Nevertheless, the belief remains widespread that profit-seeking farmers have greatly contributed to the nitrate problem (i.e. untimely nitrate) by applying too much nitrogen fertilizer, and that the excess has been washed into rivers and aquifers. We will turn our attention in Section 4.5 to what *actually* happens to the nitrogen fertilizer they apply. But first we ought briefly to reflect on the main points of the chapter so far.

Activity 4.5 *You should spend up to 15 minutes on this activity.*

Summarize the chapter so far by *writing down* what you consider to be the *most important points* (up to, say, ten) to have been made about the nitrate problem. In order to do this, you will probably have to scan the previous pages (perhaps looking for any highlighting you have done).

4.5 What really happens to nitrogen fertilizer?

If you want to trace the movements of birds or other animals you tag them. Nitrogen fertilizer can be tagged with ^{15}N, a heavy isotope of nitrogen. This behaves like ordinary nitrogen in chemical and biological terms, but can be distinguished from it by using a mass spectrometer. Activity 4.6 invites you to help design experiments in which this tagging facility is used to follow the fate of nitrogen fertilizer applied to winter wheat (this crop having been chosen because winter cereals occupy about 60% of the arable land in the UK).

Activity 4.6 *You should spend up to 15 minutes on this activity.*

Aim of the experiments

To find out where (in the soil, in the crop or elsewhere) nitrogen applied as fertilizer is found at the time of harvest.

In the course of the experiments, factors such as the amount of fertilizer applied and the timing of its application would be varied systematically. It would also be necessary to replicate each treatment in order to allow for factors beyond the experimenters' control (e.g. variation in soil quality). However, this activity is concerned with the *basic experimental design* common to all experiments and replicates.

Materials and equipment

These are listed below, but you can have the use of other items if necessary.

- Plots of winter wheat, sown ready for the experiments.
- Nitrogen fertilizer, tagged with ^{15}N, supplied as a solution in water.
- The Rothamsted '^{15}N bedstead', a device designed to ensure uniform distribution of the dissolved fertilizer.
- A sickle for harvesting the crop (the plots would be too small for a combine harvester to be used) and a small threshing machine.
- A motorized screw-auger for taking some fairly large soil samples.
- Facilities for analysing the soil, including a mass spectrometer.

Methods

Having been told the aim of the experiments and given a list of the materials and equipment available to you, your task is to think about the basic experimental design common to all the experiments and replicates. Answer the following questions and then compare your answers with the approach adopted by the Rothamsted ^{15}N team in their series of experiments, as outlined in the answer.

(a) At what time of year would it be of most relevance to apply the tagged fertilizer?

(b) What special precautions would you take when applying the tagged fertilizer?

(c) Where would you look for the ^{15}N at the time of harvest?

(d) Would you expect to recover all the ^{15}N? If not, where would you expect the 'missing' ^{15}N to go?

Tagged nitrogen which the Rothamsted experimenters were unable to account for was assumed to have been leached or lost into the atmosphere as a result of a process known as **denitrification** (Figure 4.3). Microbes need oxygen to metabolize the carbon-containing matcrials on which they feed. If insufficient gaseous oxygen is available in the soil (e.g. because it is waterlogged), denitrifying bacteria can use nitrate as an alternative source of oxygen; the nitrate is reduced to nitrous oxide (N_2O) or nitrogen gas (N_2), which are then lost into the atmosphere.

The percentages of ^{15}N found in the locations listed in the answer to part (c) of Activity 4.6 at the time of harvest varied considerably between the different Rothamsted experiments. We can therefore give you only ranges of values:

- 50–80% was found in the grain, chaff and straw of the crop.
- 10–25% was found in the roots, stubble and soil organic matter.
- 1–2% was found in the soil as N_{min}, i.e. as ammonium or nitrate.
- The proportion found in weeds varied greatly according to how well the crop had competed with the weeds for it (which in turn depended on how much nitrogen fertilizer had been applied).

Figure 4.5 shows, for each of 13 experiments in which fertilizer was applied in spring, the percentage of ^{15}N that was not found—and therefore presumed lost. Computer models were used to estimate the percentage lost by leaching in these experiments and the percentage lost by denitrification was then obtained by subtraction.

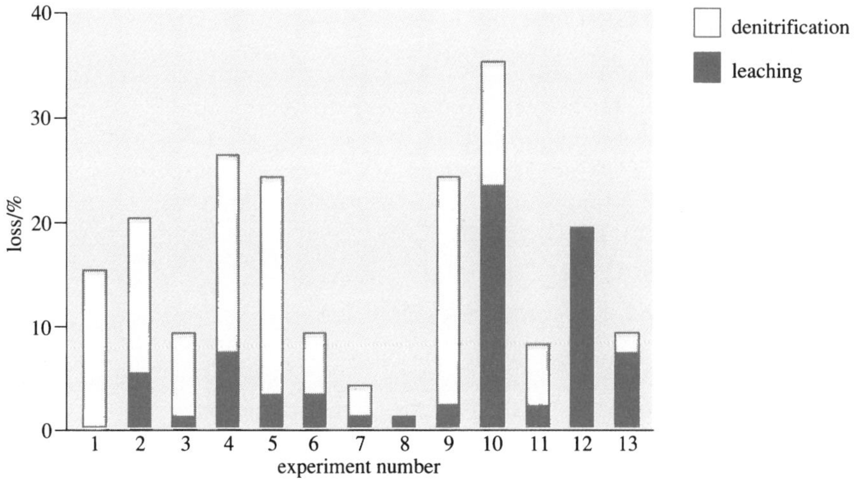

Figure 4.5 The partitioning of losses of ^{15}N between leaching and denitrification in 13 experiments in which tagged fertilizer was applied to winter wheat in spring.

▷ What were (a) the lowest and (b) the highest percentages of ^{15}N lost in these 13 experiments?

▶ (a) About 1% (in Experiment 8) and (b) about 35% (in Experiment 10).

▷ What was the average percentage of ^{15}N lost?

▶ Estimating by eye, about 15%.

▷ What were the average percentages of ^{15}N lost (a) by leaching and (b) by denitrification?

▶ Again estimating by eye, (a) about 5% of the ^{15}N was lost by leaching and (b) about 10% of the ^{15}N was lost by denitrification.

In the few experiments in which the tagged fertilizer was applied in autumn the losses were much greater (about 40%) than the losses when fertilizer was given in spring. The experimenters concluded that leaching must have played a greater part in the autumn loss than it did in the spring loss.

Question 4.1 What conclusions of relevance to the nitrate problem can be drawn from the results of the Rothamsted experiments presented above? For instance, what proportion of nitrogen from applied fertilizer ends up where intended, what proportion remains in the soil and so vulnerable to leaching, when should nitrate fertilizer *not* be applied? *Write down* your own ideas before reading our answer.

We now need to pay a little more attention to the process of denitrification, which accounted for an average of about 10% of the ^{15}N applied in spring in the Rothamsted experiments.

▷ Which two gases are released into the atmosphere as a result of denitrification?

▶ Nitrogen (N_2) and nitrous oxide (N_2O).

Since N_2 makes up 78% of the atmosphere in any case, its release presents no environmental problems. This is certainly *not* true of N_2O, which is both a more potent 'greenhouse' gas than CO_2 and one of the gases that damages the ozone layer.

So, is denitrification a 'good thing' or not?

Question 4.2 Bearing in mind that crops *need* nitrate and that microbes do what is best for *themselves*, what are the advantages and/or disadvantages of denitrification (a) as a means of cutting the amount of nitrate leaching from agricultural land and (b) as a means of removing nitrate from water in water-treatment plants (a technique currently used in the UK, France and Germany)?

4.5.1 What contributes to untimely nitrate?

The Rothamsted experiments suggest that, on average, only 6–7% of the fertilizer given to winter wheat in spring is lost directly by leaching. This is not a spectacular loss, so is it the worst that can happen? The answer, unfortunately, is 'no'. The farmer, the crop, the soil and the weather can all make things worse.

The *farmer* can put on too much nitrogen fertilizer and/or put it on at the wrong time. Both contribute to untimely nitrate. If too much nitrogen is applied, an unnecessarily large amount of nitrate is left in the soil at harvest. (As we have seen, there will inevitably be *some* nitrate as a result of microbial activity.) There is evidence from field plots of winter wheat that the amount of N_{min} in the soil at harvest remains almost constant over quite a wide range of application rates of nitrogen fertilizer before starting to climb quite steeply (Figure 4.6).

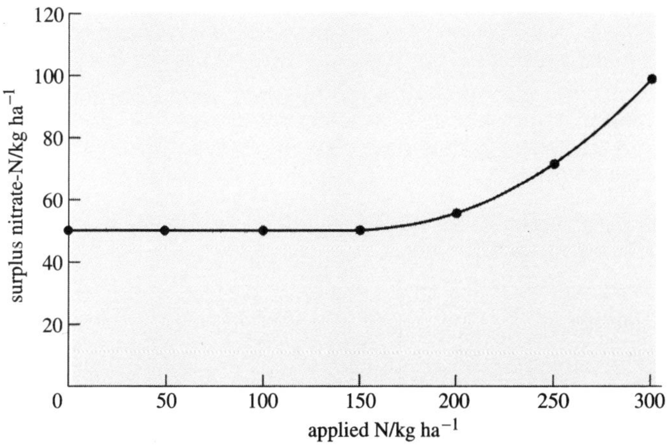

Figure 4.6 A 'surplus nitrate' curve for winter wheat.

▷ What is implied by this 'surplus nitrate' curve?

▶ That, up to a 'point of satisfaction', the crop can use extra nitrogen but that further nitrogen is likely to contribute to the nitrate problem.

This 'point of satisfaction' could be considered an *environmental optimum* for application of nitrogen fertilizer. There will also be an *economic optimum*, beyond which the costs of applying more fertilizer outweigh the returns for doing so. Fortunately, the environmental and economic optima for winter cereal crops usually coincide. Furthermore, the cuts in crop prices in the recent Common Agricultural Policy (CAP) reforms will lower the economic optimum.

▷ What formerly common practice that we have already discussed usually led to increased leaching of nitrogen, but from which the crop benefitted little?

▶ Application of nitrogen fertilizer in autumn to winter wheat and other autumn-sown crops to 'help them through the winter'.

The *crop* can contribute to the problem by not taking up the nitrogen it is given. A crop which suffers drought or disease, and which therefore does not grow to its expected potential, can leave appreciable amounts of nitrate in the soil at harvest.

Through its texture, *soil* influences the loss of fertilizer nitrate. Nitrate is more likely to be leached from a sandy soil than from a silt or clay soil.

With respect to the *weather*, suffice it to say that the largest loss of ^{15}N in the Rothamsted experiments occurred when 114 mm of rain fell in the first three weeks after the fertilizer was applied.

Clearly, a large number of interacting factors contribute to the agricultural side of the nitrate problem. These include inputs of nitrate by the farmer and by soil microbes, uptake of nitrate by the crops and its washing out by rainfall.

4.6 The Nitrate Sensitive Areas schemes

The **Nitrate Sensitive Areas (NSAs)** schemes were brought in by the Ministry of Agriculture, Food and Fisheries (MAFF) to help protect water supplies that are vulnerable to pollution by nitrate for geological reasons.

Farmers in ten NSAs have been invited to join one of two schemes. The *Basic Scheme* is intended to decrease nitrate leaching, broadly within the context of existing agricultural practices. The *Premium Scheme* involves more fundamental changes, such as putting arable land down to grass. Although participation in the schemes is voluntary at present (which farmers prefer), water suppliers seem keen for it to be made compulsory.

The Basic Scheme involves:

- Limiting applications of nitrogen fertilizer (chemical or organic) and constraining applications to within certain times.
- Requiring the farmer to sow a crop, or at least a 'catch' crop, so that the land is not bare in autumn.
- Limiting the ploughing of grassland; only leys (i.e. grassland that is in rotation with arable crops) may be ploughed.
- Retaining hedgerows and woodland or replacing them by equivalent features.

- Drawing up plans for pig and poultry units to show that they can store, handle and transport manure in a way that enables it to be spread according to the requirements of the scheme.

The Premium Scheme involves converting arable land to grassland. Its four options are:

- To leave the grass unfertilized and ungrazed.
- To have the land grazed, but apply no fertilizer to it.
- To apply a limited amount of fertilizer and have the option of grazing.
- To convert arable land to grassland with woodland.

Farmers are paid according to the NSA in which they farm, the proportion of their land submitted under the schemes and the extent of the measures that they undertake. ADAS is setting up an extensive programme to find out how effective the schemes are in restricting nitrate losses from farmland. Other organizations also seem likely to make evaluations, and the results will be used to test computer models that simulate nitrate losses from farmland to aquifers and surface waters. Once validated, the models can be used to predict the likely future progress of the NSAs.

Activity 4.7 *You should spend up to 20 minutes on this activity.*

Explain, in not more than 200 words, the scientific reasoning underlying the NSAs schemes.

4.7 Who else contributes to the nitrate problem?

Contrary to what you sometimes hear, the farming community has not created the nitrate problem all on its own. I doubt you would be surprised to learn that politicians have played their part! As recently as 1975, a government White Paper stated that 'The government takes the view that a continuing expansion of food production in Britain will be in the national interest'. That continuing expansion involved the use of increasing amounts of nitrogen fertilizer, which must have built up the levels of mineralizable nitrogen in soils—even if it did not contribute directly to nitrate leaching. And what about Euro-politicians and the famous CAP, which assured farmers of a market and a consistent price for their produce even when there was a glut of it? The inevitable result was excessive production, obtained in part by increased use of nitrogen fertilizer.

What about your own pension fund? Surely that could not have contributed to the problem? Well, about 20 years ago agricultural land suddenly began to look a good investment—probably because of the CAP. Large corporate bodies (including pension funds) began to invest in it, and its price went up. Agricultural land prices started to increase rapidly from about 1970 and roughly doubled between 1975 and 1980. The corporate investors, and also farmers who had borrowed considerable sums of money, needed a good return on their investment. In any case, interest rates have hardly been static in recent years. This has been bad enough for those with a mortgage on a 'semi'—imagine what it is like paying for a large farm! Crops therefore had to be 'squeezed' to yield as profitably as possible. Bring on the nitrogen fertilizer! If you are 'squeezing' a wheat crop, for example, you might carry on getting a financially worthwhile return from nitrogen fertilizer even after you go past the point at which all the fertilizer is used efficiently (Figure 4.6). The rest you can work out for yourself.

Water companies are, of course, keen to keep down nitrate concentrations. However, their main business is selling water, which means it has to be taken out of rivers and aquifers. The nitrate concentration in a river or aquifer increases not only when more nitrate enters it, but also when more water is taken out (because there is less water to dilute the incoming nitrate). To some extent therefore, the increased extraction of water in recent years is likely to have contributed to the nitrate problem.

Finally, what about *you*? Like soil microbes and farm animals, you too put nitrate into the environment in the everyday business of living. You are also a consumer—of both water and food—and presumably prefer your food to be as cheap as possible. Nitrogen fertilizer has been one of the factors that has helped to keep food cheap. If you eat meat and consume dairy produce, recall how inefficient farm animals are in converting nitrogen fertilizer into human food. In the final activity of this chapter, we would like you to consider *your own role* in the nitrate problem and what you propose to do about it.

Activity 4.8 *You should spend up to 20 minutes on this activity.*

Although farmers get most of the blame for the nitrate problem, you are responsible for it too. So, what are you going to do about it? Try writing down a *personal* strategy for combating the nitrate problem. Think carefully about your home, your job, your recreational activities, your personal habits and all your day-to-day affairs, including what you eat. Identify the points at which you contribute to the nitrate problem and work out how you could reduce your contribution at these points.

Summary of Chapter 4

1 The evidence linking nitrate in clean water supplies to either methaemoglobin-aemia or stomach cancer is fairly weak, except where nitrate concentrations are really excessive ($>200\,\mathrm{mg\,l^{-1}}$). However, it does seem that excess nitrate in rivers and lakes contributes to eutrophication of the water (in the presence of sufficient phosphate, which is largely derived from discharges from sewage works).

2 Nitrogen gas from the air is fixed into ammonium and ultimately nitrate by free-living algae and bacteria in the nodules of leguminous plants, as well as by the commercial manufacturing of fertilizers.

3 Much of the world's soil-borne nitrogen is stored in a relatively inert form in dead organic matter.

4 The biomass of the soil (in particular, soil microbes) releases nitrates from this organic matter. This happens on a large scale when old grassland is ploughed up. It happens particularly in the autumn, when the soil is still warm from summer but becoming more moist; thus it is very important that either a winter or a 'catch' crop is available to use this nitrate.

5 Farming based on animal husbandry is particularly inefficient in its utilization of nitrogen from fertilizer applied to grassland.

6 Untimely nitrate is nitrate present in the soil at a time when it cannot be utilized by a crop. Such nitrate is likely to be leached beyond reach of the crop's roots and end up in rivers, lakes and aquifers.

7 The Rothamsted ^{15}N experiments suggest that much of the fertilizer nitrogen applied to winter wheat in spring ends up in the harvested parts of the crop. Most of the remainder ends up in organic material in the soil, where it is safe from leaching.

Only 1–2% of the nitrogen is left as mineral nitrogen (ammonium and nitrate) in the soil, and therefore vulnerable to leaching. On average 15% of the nitrogen was lost, one-third of which was probably leached and two-thirds converted into N_2 and N_2O through denitrification by soil microbes. Nitrogen fertilizer applied in the autumn was leached from the soil to a much greater extent than that applied in the spring.

8 Denitrification may be useful in water-treatment plants, where it can be controlled to minimize emission of environmentally harmful N_2O.

9 The Nitrate Sensitive Areas schemes are designed to protect water supplies that are vulnerable to pollution by nitrate for geological reasons; they are based on the scientific principles discussed in this chapter.

10 All individuals and organizations contribute to the nitrate problem; it is therefore incumbent upon us all to support strategies to combat the problem.

5 Bovine spongiform encephalopathy (BSE) (3hrs)

Of all the chapters in the book, this is the one for which it is most certain that new information will come to light during the lifetime of the Course. We hope that you will be sufficiently interested to follow such developments yourself, building on the understanding of the disease you have gained from studying this chapter, which is based on knowledge as it stood in July 1996. Note, however, that this chapter is primarily concerned with BSE in cattle *not* with CJD in humans, except in so far as there *may* be a link between the two conditions.

5.1 Introduction

Following the diagnosis, during 1994 and 1995, of ten cases of a new variant of **Creutzfeldt–Jakob disease (CJD)** in people under 42 years of age, the Secretary of State for Health stated in the House of Commons on 20 March 1996:

> *'there remains no scientific proof that BSE [bovine spongiform encephalopathy] can be transmitted to man by beef, but the [scientific advisory] committee have concluded the most likely explanation at present is that these cases are linked to exposure to BSE before the introduction of the specified offal ban in 1989.'*

Previously it had been believed that transmission of the disease between species would not occur because of the so-called 'species barrier' (Section 5.4.2). However, ten years after BSE was first diagnosed in Britain, evidence had been found to support the possibility that BSE had 'jumped species' from cattle (*Bos taurus*) to humans (*Homo sapiens*).

The public did not believe the assurance by the Secretary of State that 'British beef can be eaten with confidence'. Retail sales immediately dropped dramatically (but began to recover when the price fell!). Within days, a leading fast-food chain announced that it would stop using British beef in beefburgers and the European Union (EU) imposed a world-wide ban on the export of British beef and beef products.

Question 5.1 Following the Secretary of State's statement there was criticism that the government had been reluctant to 'go public' with the findings. Why might there have been this reluctance?

The beef industry is very important in Britain. In 1995, household purchases of beef totalled about £4 billion and exports of 242 000 tonnes were worth £520 million. Apart from the 120 000 beef and dairy farmers, workers in abattoirs and industries processing animal by-products would have been adversely affected by any loss of confidence in the industry. But how did it all start?

In early 1985, UK vets observed a few cases of cattle with unusual neurological symptoms in at least three herds throughout the country, but examinations of brain cells of affected animals were not carried out until more than a year later. In November 1986, the first case of the *new* disease **bovine spongiform encephalopathy (BSE)** (in other words 'spongy brain disease in cattle') was confirmed. Initially, the occurrence was sporadic. In 1987, there were fewer than 100 cases per month, but an epidemic developed as the number of cases throughout the country grew (Figure 5.1).

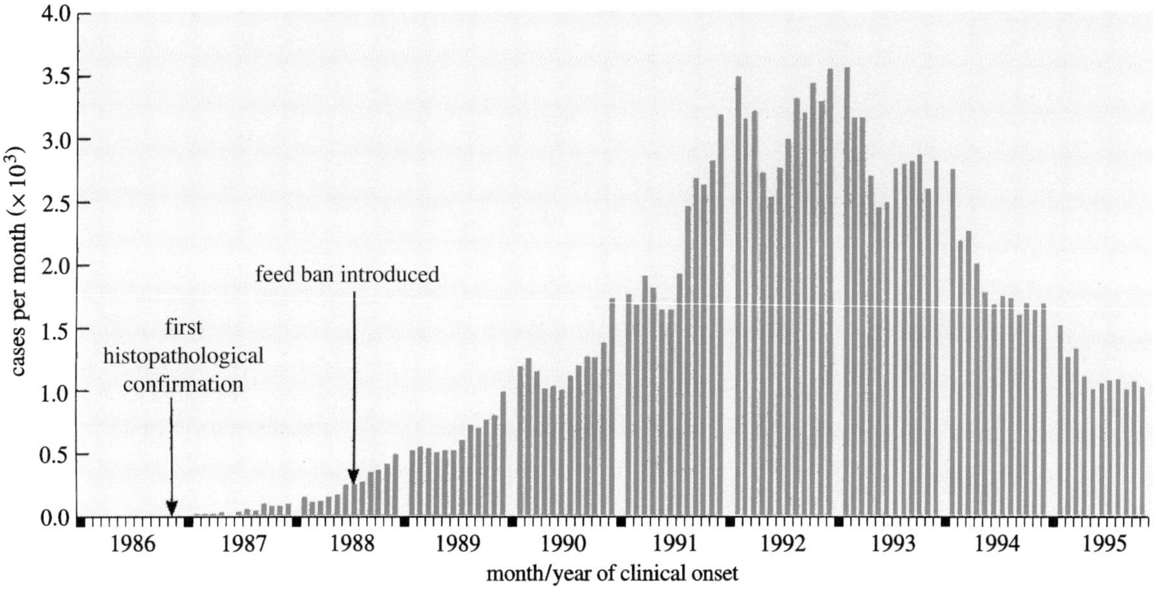

Figure 5.1 The number of confirmed cases of BSE shown by month and year of clinical onset, i.e. when the symptoms first became recognisable. Histopathology is the microscopic examination of diseased cells.

▷ How many cases were recorded per month (a) at the height of the epidemic and (b) in late 1995?

▶ (a) More than 3 500 cases were recorded per month at the height of the epidemic in the winter of 1992/93. (b) By late 1995, there were still about 1 000 cases per month being recorded.

By the end of 1995, 155 000 cases had been confirmed from 32 906 farms in the UK.

Farmers describe symptoms of the disease as nervousness or altered behaviour or temperament. Typically, animals start to kick during milking and exhibit difficult and awkward movements — hence the inaccurate but commonly used name 'mad cow disease'. They also lose weight and milk yields decline. On post-mortem examination, degenerative symptoms in the central nervous system are visible under the light microscope. Fluid-filled cavities — vacuoles — are present in brain cells which give them a spongy appearance and, in some cases, nerve cells appear to be shrunken and dying (Figure 5.2 overleaf).

Before the appearance of BSE, six **transmissible spongiform encephalopathy** (**TSE**) diseases had been recognised world-wide, three in humans and three in other animals (Table 5.1). All affect the central nervous system and have similar effects on the brain to those described above for BSE. The diseases are associated with abnormalities in a single cellular protein — designated the prion protein — whose role will be discussed further in Section 5.4. An important characteristic of TSE diseases is the long incubation time — measured in months and often years — between infection and the onset of recognisable symptoms. It is also important to note that diagnostic tests for TSEs in live animals have only recently (1996) been developed and their reliability has still to be determined.

Table 5.1 Transmissible spongiform encephalopathy diseases recognised before the appearance of BSE.

TSE disease	Host
kuru	humans
Creutzfeldt–Jakob disease (CJD)	humans
Gertmann–Straussler syndrome	humans
scrapie	sheep and goats
chronic wasting disease	mule, deer and elk
transmissible mink encephalopathy	farmed mink

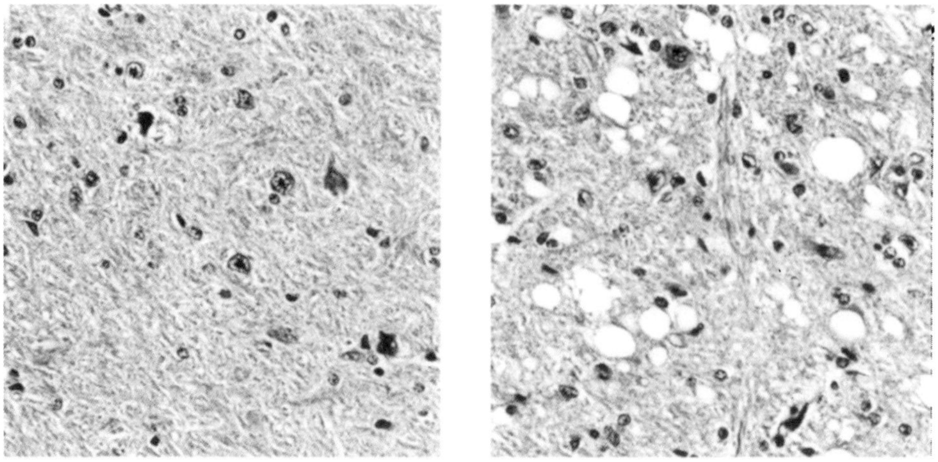

Figure 5.2 Brain cells from (a) a healthy cow and (b) a cow affected by BSE – note the prominent vacuoles. (Magnification × 240)

Such diseases in humans are rare. World-wide, CJD is the most common with about one case per million people. There are three types of CJD: familial (10–15% of cases) caused by genetic mutation; cases caused by transmission of disease from infected material during medical treatment (<1% of cases), e.g. from contaminated human growth hormone; and sporadic (85%–90% of cases) from unknown causes. BSE is linked to this last type of CJD. Of the animal diseases, **scrapie** is the most widespread as it occurs in sheep and goats in many countries throughout the world and has been known for many years.

Three issues surrounding BSE will be addressed in this chapter. The first is where BSE came from; this will be investigated by considering the epidemiology of the disease (Section 5.3). The second is what causes BSE; researchers have worked for many years to isolate the agent responsible with no positive result (Section 5.4). The third is the question of whether humans can contract CJD from eating beef contaminated with BSE (Section 5.5).

But to be able to interpret the epidemiological data, it is first necessary to have a basic understanding of how cattle are farmed in the UK.

5.2 Cattle farming systems

▷ What are the main products from cattle farming?

▶ Meat (beef) and dairy (milk) products.

On the one hand, cattle are required to produce milk for dairy products and, on the other, to produce muscle for beef. Over the centuries, therefore, cattle have been bred primarily for one or other of these functions, and have been farmed under different production systems. What follows is a highly simplified description of each system and the link between them.

In UK dairy herds, the herd size can vary from fewer than ten animals to more than 200. Some 90% of female cattle belong to the Holstein Friesian breed and 5% to the Channel Island and Ayrshire breeds. Dairy cattle are housed indoors for about seven months of the year, are regularly treated with medication (e.g. antibiotics) and are handled twice each day during milking. Each year, cows of two years old or more, are mated and produce a calf. Because breeding from good stock is so important, most cows undergo artificial insemination with semen from one of relatively few bulls with known pedigree. As a result, one bull may sire tens of thousands of cattle in a year. Soon after birth, calves are taken from their mothers and fed on reconstituted dried milk and protein-rich concentrated feed (concentrates). Adults are also fed on concentrates at times when their energy demand exceeds that available from grass and its products. The milk from the cows is harvested for human consumption. At about five and a half years of age, when the milk yield starts to decline, dairy cows are slaughtered and their meat was, until recently, used in cheaper meat products such as pies, burgers and sausages.

Breeds of beef cattle include the Aberdeen Angus, Galloway, Highland, Hereford and Charolais. They can be farmed in very intensive systems but the majority live outside in beef suckler herds for most of the year, receive little medication and feed on grass, hay or silage. Beef cows also undergo artificial insemination, but the calves suckle from their mother for several months and are not fed on concentrates. Both male and female calves are reared for one to two years before they are slaughtered for meat.

There is an important link between the dairy and beef herds because only 20% of the calves produced by dairy cows need to be kept in the dairy herd to provide replacements. The remaining 80% of calves are sold for rearing as beef cattle or for veal. Because of this, beef sires are frequently used for artificial insemination of dairy cows to ensure that the resulting offspring are genetically suitable for beef production.

5.3 Investigating the disease

This section gives some of the results from the early epidemiological study (Section 5.3.1) which led to the development of a hypothesis about the origin of the disease (Section 5.3.2). The measures introduced to control BSE in the UK cattle herd are presented (Section 5.3.3) and the issue of direct transmission of BSE between animals within a herd is discussed (Section 5.3.4).

5.3.1 The epidemiological evidence

From the beginning of the BSE scare, the progress of BSE was monitored by scientists at the Central Veterinary Laboratory. When it became clear that the number of cases was increasing into possibly epidemic proportions, it was considered most important to investigate which animals were affected, how BSE was spreading and where it came from. To do this type of investigation it is necessary to study the disease in *populations*, and take what is called an **epidemiological approach**, rather than making a study of diseased *individuals*.

▷ By what categories could cattle be sensibly grouped?

▶ By age, sex, 'occupation' (i.e. dairy or beef), breed, herd size and geographic location.

Information on the incidence of BSE according to *some* of these categories has been selected for interpretation in Activity 5.1. The information is mainly taken from a scientific paper*, published in 1988, which presented the results of the early epidemiological studies and which enabled scientists to formulate a hypothesis about the origin of the disease.

A **hypothesis** is a suggested explanation for the factual evidence that has been collected. Once formulated, the hypothesis can be used as the basis of further investigation. It is important to remember a basic scientific principle — that it is impossible to prove a hypothesis, but relatively easy to disprove one. For example, observing 1 000 white swans does not prove the hypothesis that all swans are white; but seeing one black swan would disprove the hypothesis. What one should do, therefore, is to look for data that would disprove a hypothesis rather than data that would prove it.

* Wilesmith, J. W., Wells, G. A. H., Cranwell, M. P. and Ryan, J. B. M. (1988) Bovine spongiform encephalopathy: epidemiological studies, *The Veterinary Record*, **123**, pp. 638–644.

Activity 5.1 *You should spend up to 15 minutes on this activity.*

Read the answer to each main part (a–e) of the activity before continuing to the next part.

(a) Figure 5.3 shows the age distribution of cases of BSE reported in 1988.

(i) What do you think the authors of the original scientific paper meant by 'age-specific incidence' of the disease?

(ii) What can you conclude from Figure 5.3 alone?

(iii) How does your knowledge of cattle production systems and TSE diseases affect your conclusion?

(b) In a sample of 192 cases of BSE in the UK, two were in male cattle and 190 were in female cattle.

(i) From your knowledge of cattle production systems, how would you explain these data? Read the answer to this part before going on.

(ii) From the numbers of male and female cattle given in the answer to (b)(i), what is the percentage incidence in males and females?

(c) Table 5.2 shows the incidence of BSE in beef suckler and dairy cows from 1985 to 1988 in relation to the number of cattle in the national herd.

(i) Calculate the percentage incidence of BSE in the national dairy herd and enter your value in Table 5.2.

(ii) What does the completed Table 5.2 suggest about the effect of herd type on the incidence of BSE?

Table 5.2 The incidence of BSE in beef suckler and dairy cows from 1985 to 1988 in relation to the number of cattle in the national herd.

	No. of cases	No. in national herd	Incidence/%
beef suckler cows	14	880 000	0.002
dairy cows	696	2 324 000	

(d) Figure 5.4 shows the sources of cattle with BSE in affected herds.

(i) What were the main sources of BSE cases in dairy and beef suckler herds?

(ii) From your knowledge of the production systems, how could the cases in beef suckler herds be explained?

(e) Figure 5.5 shows the incidence of BSE-affected dairy herds by county/region for the period 1985 to 1988.

(i) For what reason do you think the authors chose to plot percentage incidence rather than just the number of herds affected per county or region?

(ii) What patterns, if any, can you distinguish in the data and what anomalies do there appear to be?

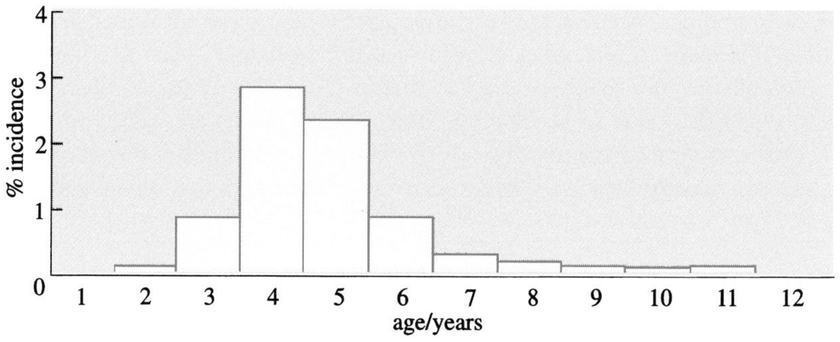

Figure 5.3 Age-specific incidence of BSE-affected herds in 1988.

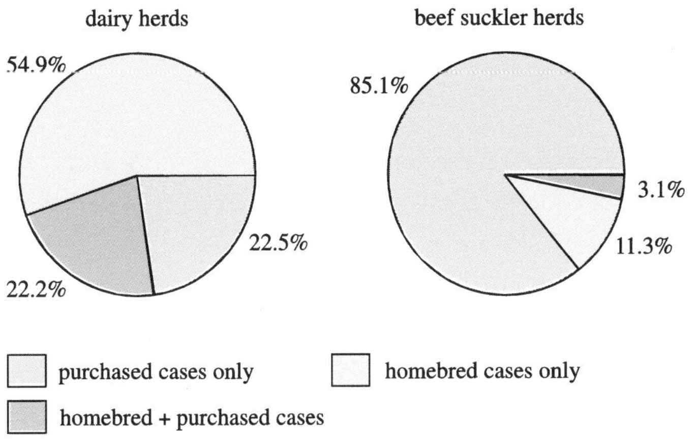

Figure 5.4 The sources of BSE cases in affected herds. Cases could be in either homebred cattle (i.e. in calves which were born on the farm), or in cattle which had been purchased from another producer. The pie diagrams show the percentage of herds with cases in homebred cattle only, in purchased cattle only and in both homebred and purchased cattle.

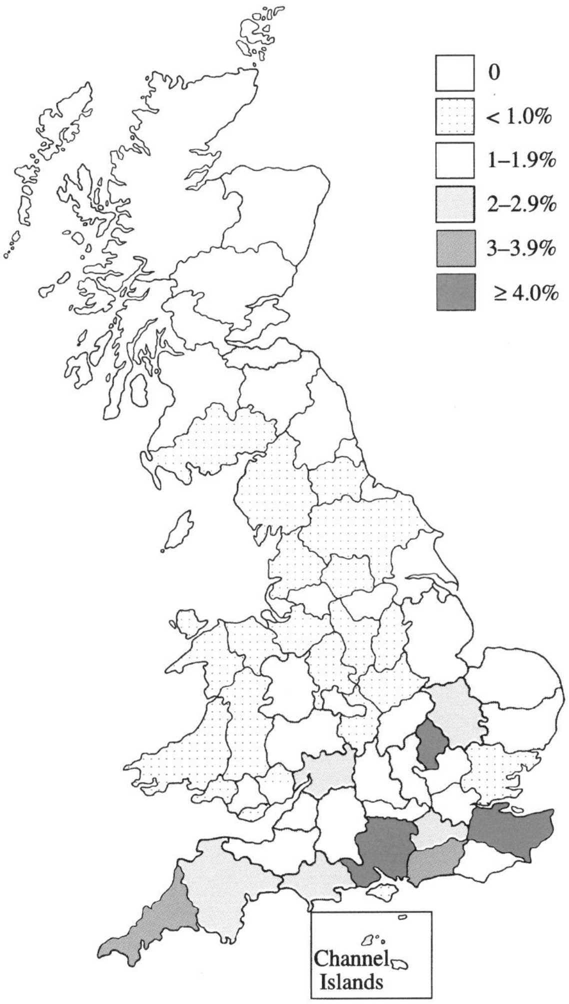

Figure 5.5 Incidence of BSE-affected dairy herds from 1985 to 1988 as a percentage of the total number of herds by county/region.

The information from the epidemiological study is summarized below:

- BSE, while not confined to any one age group, occurs in older breeding animals, perhaps as a result of having a long incubation period.
- Both calves and adults had been exposed to whatever causes BSE but the probability of exposure was 30 times greater for calves than for adults.
- Exposure commenced in the winter of 1981/82 and continued until at least the end of 1984.
- Incubation periods ranged from 2.5 to at least 8 years, although they were biased towards shorter periods.
- The incidence of BSE was similar in both sexes of cattle.
- The incidence of BSE was much greater in dairy cattle than beef cattle.
- Cases of BSE in beef herds were probably in cattle purchased from dairy farms.
- No distinct geographic focus of the disease could be identified.

▷ Which animals presumably had the highest level of exposure to whatever agent caused BSE?

▶ Calves born into dairy herds.

▷ Are there any differences in the way dairy and beef calves are treated that could explain the epidemiological information?

▶ Calves on dairy farms are taken from their mothers and are fed on reconstituted dried milk and protein-rich concentrates. They are also exposed to medications and agricultural chemicals. In contrast, calves born on beef farms suckle their mother's milk and have a lower level of exposure to medications and chemicals.

In this study, it was considered unlikely that BSE came from chemicals used on farms because, after analyzing the results of a detailed questionnaire sent to farmers, it proved impossible to identify any common chemical link between all the infected farms. Therefore, the hypothesis that we will now test is that 'BSE came from consumption of concentrated feed by dairy calves'.

5.3.2 The cattle-feed hypothesis

Concentrated cattle feeds are mixtures of several ingredients which are ground up, mixed thoroughly and then pressed into pellets. Concentrates are produced to nutritional specifications under commercial conditions and a wide variety of ingredients is used in order to keep costs down. Could one of these ingredients be the source of infection?

The majority of the ingredients in concentrates are of plant origin but until recently one was of animal origin. **Meat and bone meal (MBM)** is a valuable source of protein derived from parts of all animals killed in abattoirs; these parts include the feet, brains, intestines, lungs and excess fat, and are all now rejected for human consumption. In a process known as **rendering**, the materials are treated to separate the fat from the residual solid material. The latter is ground up and sold as MBM.

Was it possible that the agent that causes scrapie in sheep (Table 5.1) was present in the MBM added to concentrated feeds and that this agent crossed the species barrier to infect cattle? There are two arguments against this hypothesis.

- Only a small proportion of calves fed concentrates have developed the disease.
- Scrapie has been present in the sheep population for more than 200 years and animal products have been used in cattle feed for several decades, so why was BSE recognised only since 1985?

So, can the hypothesis be rejected because there are powerful arguments against it? Or, were there any other relevant events which occurred around 1980–82 which may explain the discrepancies?

At about this time (1980), the government allowed a relaxation in the regulations controlling the rendering process. Processing had previously involved first heat-treating the waste animal material to remove water and fat, and then using a hydrocarbon solvent to dissolve residual fat. The solvent was then removed with steam at temperatures of between 100 and 120 °C. Processing plants varied considerably in their procedures but in all it took several hours to complete the hydrocarbon extraction—a wet process applied in a low fat environment. After the change in regulations, dry heat was applied to remove water and fat from the material, and the hydrocarbon extraction was phased out (Figure 5.6). It is now thought that the new procedures were less effective at deactivating any scrapie agent present in MBM. By 1988, only two of the 46 plants in Great Britain were still using solvent extraction and these were in Scotland.

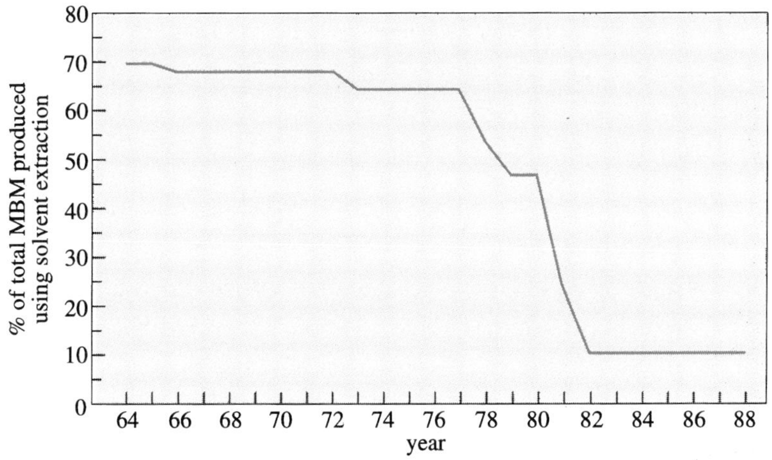

Figure 5.6 The proportion of meat and bone meal produced from 1964 to 1988 by rendering plants using solvent extraction.

▷ Cattle feed is produced in many rendering plants around the country which supply farms in their immediate locality. Does the fact that the only two plants still using the solvent extraction process in 1988 were in Scotland relate to the geographic incidence of the disease shown in Figure 5.5?

▶ Yes, there were no cases of BSE in cattle in central and northern Scotland during 1985 to 1988. In southern Scotland up to 1.9% of herds were affected. In some parts of England more than 4.0% of herds had cases of BSE.

In 1988, an hypothesis emerged that was generally accepted as likely to be true: BSE came from consumption of feed concentrates which probably contained scrapie agents in MBM derived from infected sheep in abattoirs. The change from the solvent extraction process during rendering of animal waste probably increased the level of contamination of the MBM with the agent.

Other hypotheses exist but are not widely accepted by the scientific community. For example, the one that perhaps has received the most publicity: BSE is caused by treatment of cattle with organophosphate insecticides. The organophosphates are rubbed into the spines of cattle to control warble fly and can, under some circumstances, adversely affect the nervous system.

5.3.3 Control of animal health

In 1988, once the epidemiological data were assessed, the government introduced legislation to make BSE a notifiable disease, to ban inclusion of *ruminant* protein in ruminant feed and to pay compensation for slaughter of suspected cases of BSE.

Investigations were made to identify which tissues of cattle with BSE were infective. The test involved laboratory mice. Homogenates from a wide range of bovine tissues were given to mice—either orally (i.e. by mouth) or by injection into the brain or body cavity. The mice are said to be *challenged* by giving such a homogenate; the test is an example of a bioassay because it is carried out on a living organism. To date, only brain tissue has proved infective to mice challenged by the oral route; but tissue from the brain, parts of the spinal cord and the retina from the eye have proved infectious when injected into the brain.

These data, together with studies on scrapie-infected sheep, led to the identification of **specified bovine offals (SBOs)** which are considered the most likely organs to contain the infective agent. These now include the head, spinal cord, thymus, tonsils, spleen and intestine. (They are currently referred to as specified bovine materials (SBMs) as the head is not classed as an offal.)

A ban on the inclusion of SBOs in *any* animal feed was introduced in September 1990 following the experimental transmission of BSE to a pig and the diagnosis of several cases of TSE diseases in wild cats and ruminants in zoos—presumably the result of consuming infected concentrates. Table 5.3 below gives a chronology of animal health control measures.

Table 5.3 Key dates and control measures introduced to protect animal health.

Date	Control measure
Nov 1986	BSE first confirmed in Great Britain.
Jun 1988	BSE made a notifiable disease.
Jul 1988	Ban on use of ruminant protein in ruminant feed.
Aug 1988	Slaughter of infected animals, 50% compensation for confirmed cases of BSE and 100% for negative cases.
Feb 1990	100% compensation for confirmed cases of BSE.
Sep 1990	Ban on the use of SBOs in any animal feed.
Oct 1990	Cattle breeding and movement records must be kept for 10 years.
Nov 1991	Ban on the use of MBM from SBOs in fertilizers.
Apr 1995	Requirement to stain SBOs blue.
Aug 1995	Existing rules on SBOs consolidated and streamlined.
Mar 1996	Ban on the inclusion of mammalian MBM in any feed for farm animals, including horses and farmed fish.
Apr 1996	Slaughter of cattle over 30 months of age in specially licensed plants and disposal outside the human and animal food-chains.
May 1996	Stricter rules on the handling of SBOs in abattoirs.
Jun 1996	UK government agrees to slaughter more cattle said to be at special risk of developing BSE.

▷ If animal feed was the only source of infection, how would the annual incidence of the disease in cattle be expected to change in the years following the ruminant feed ban?

▶ Because of the long incubation period of the disease, with the majority of cases in three- to five-year-old cattle, the number of cases would be expected to rise until about 1991 and then begin to drop. No cattle born after the introduction of the ruminant feed ban (July 1988) should have become infected with BSE.

Figure 5.1 shows what actually happened; the epidemic did not peak until 1992/93. By December 1995, 23 148 cases (15% of all cases) of BSE had been confirmed in cattle born after the ban. Most of these were born in the period immediately after the introduction of the ban.

▷ Why might the number of cases have remained high during 1992 and 1993?

▶ The control measures may not have been fully effective and there may also have been direct transmission—that is between animals in the herd or from parent to calf.

It is true that the regulations were not fully effective. Some farmers illegally continued to use stocks of feed bought before the ban and probably sold on cattle suspected of being BSE infected. Spot checks in abattoirs have also demonstrated non-compliance with the regulations. Since 1988, a number of control measures have been introduced to further reduce the possibility of SBOs entering the animal food-chain.

In April 1996, following the announcement of a possible link between BSE and CJD in humans (Section 5.1), further measures were taken to eradicate BSE in the United Kingdom *and* to have the world-wide ban on the sale of British beef and beef products imposed by the EU lifted. The objectives of the measures were threefold:

- to protect consumers against any risk that BSE may be transmissible to humans;
- to eradicate BSE in the UK cattle herd;
- to prevent transmission to other animal species.

Legislation was introduced to completely exclude mammalian MBM from all animal feed, to slaughter and destroy all cattle which were suspected of having BSE, and to remove and destroy all SBOs from healthy cattle under stricter conditions than before.

Despite these measures, at the time of writing, the world-wide ban has only been partially lifted.

5.3.4 Direct transmission of BSE between cattle

In **direct transmission**, animals infect each other directly without an intermediate step in the transfer of disease. It can occur horizontally, that is between cattle within a herd, or vertically between parent and offspring. **Indirect transmission** is where disease is passed from animal to animal via an intermediate step, which in the case of BSE is concentrated feed.

The mode of transmission is important because it is a factor which determines how a disease is maintained within a population and hence how it might be controlled. For BSE, the question of transmission has been debated for several years, but the long incubation period and the lack of a *reliable* diagnostic test for live animals has hampered its resolution. There are, however, three sources of indirect evidence.

- The percentage of cases of BSE within affected herds has dropped from 2.72% in 1992 to 2.06% in 1995.

▷ Do these data support the hypothesis for direct transmission?

▶ No, if direct transmission is a factor the percentage of cases in an affected herd would be expected to be greater than this as the disease spread from animal to animal within the herd.

- The age distribution of the percentage incidence of BSE in infected herds changed between 1992 and 1995 (Figure 5.7 overleaf).

▷ Do the data in Figure 5.7 support the hypothesis for direct transmission from parent to calf?

▶ No, if direct transmission occurs, a shift towards older animals of the type shown in Figure 5.7 would not be expected.

- When the *observed* annual incidence of BSE in the offspring of confirmed cases was compared with the *expected* number of cases from contaminated concentrates alone, in no year did the actual incidence exceed the expected (Figure 5.8). This suggests that the epidemic can be explained by the cattle-feed hypothesis.

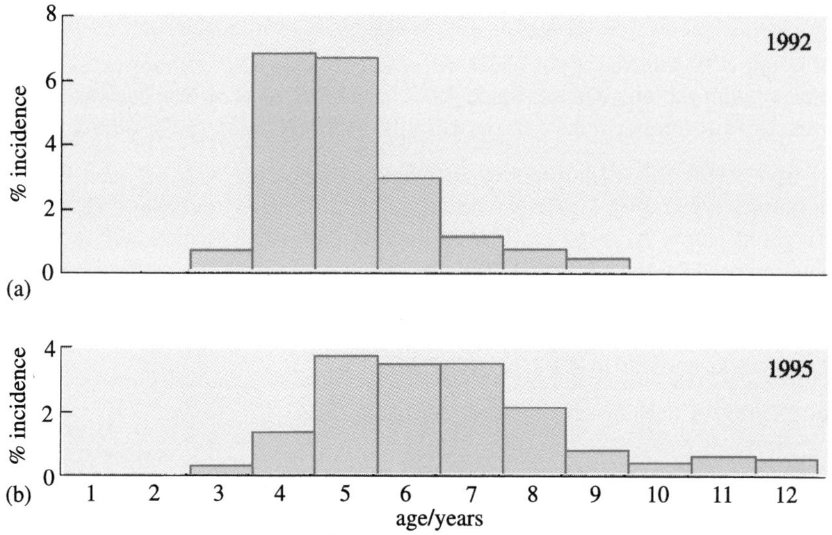

Figure 5.7 Age-specific incidence of BSE-affected herds in (a) 1992 and (b) 1995.

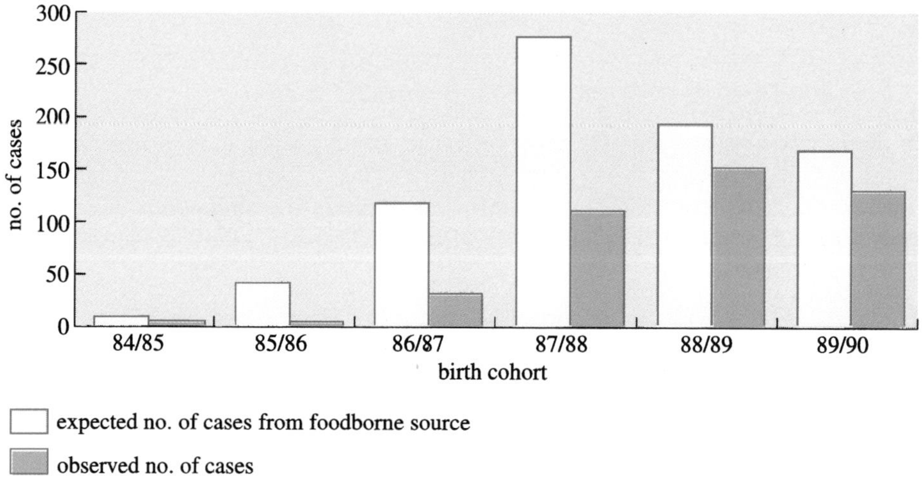

Figure 5.8 The expected incidence of cases of BSE in offspring of confirmed cases, calculated by assuming contaminated feed was the only source of infection, compared to the observed incidence of cases of BSE in offspring of confirmed cases.

Although these sources of evidence cannot prove that direct transmission does not occur, they do suggest that if it does it is at a low level.

Bull to calf transmission can be elimated because, in mouse bioassays, infectivity has never been detected in semen from BSE-affected bulls. The risk of vertical transmission from cow to calf is currently being investigated in a long term study of offspring from 315 confirmed cases of BSE and from 315 healthy cows born during the 1988/89 calving season. The experiment has been carried out 'blind', that is staff carrying out the investigation are unaware of the BSE status of the cows. Although the full results of this study will not be available until 1997, a preliminary announcement in mid-1996 indicated that there is *some* cow to calf transmission. However, how this transmission occurs is currently unknown.

5.4 What is the agent that causes BSE?

So far in this chapter, whatever it is that causes BSE and scrapie, has been described as an 'agent'. Here the nature of the agent is addressed. In much of the section, examples will be taken from studies on scrapie as this has been the most widespread and economically important of the TSE diseases and has been studied for many years.

▷ What are the causes of disease?

▶ Biological agents (such as bacteria, fungi, viruses, parasitic worms and insects) and chemical agents. There is also the possibility of genetic transmission, as in familial CJD (Section 5.1).

As discussed in Section 5.3, investigations on BSE found no association with any of the chemical agents used on farms nor was there any evidence of genetic transmission. There was also no microscopic evidence of invasion of the brain by parasitic worms or insects. Studies on scrapie demonstrated that the agent could be filtered through a mesh too small to pass bacteria, fungi or even some viruses. So at first, the agent was thought to be a very small virus. Viruses are composed only of nucleic acids (DNA or RNA) surrounded by a layer of protein.

▷ What is the function of the nucleic acids?

▶ Nucleic acids hold the genetic information which enables the virus to reproduce itself.

Tests on purified preparations of scrapie-infected brains gave the results shown in Table 5.4.

Table 5.4 The effects of ultraviolet (uv) radiation and enzymes on the degree of infectivity of the scrapie agent.

Treatment	Effect
uv radiation	Infectivity unaffected by wavelengths which inactivate nucleic acids, but destroyed at shorter wavelengths.
enzymes	Infectivity decreased by enzymes which attack proteins (**proteases**) but not by enzymes which attack nucleic acids, carbohydrates or fats.

▷ Is there evidence from the results in Table 5.4 that the scrapie agent is composed of proteins and/or nucleic acids?

▶ There is evidence that the agent contains proteins as infectivity decreases with exposure to proteases. However, there is no evidence that the agent is composed of nucleic acids as treatments which destroy nucleic acids do not decrease the infectivity of the agent. Therefore, the agent cannot be a virus.

Several decades of exhaustive analyses had failed to identify any foreign nucleic acid in samples taken from TSE-diseased organisms. So, in 1982, Stanley Prusiner, a biochemist working at the University of California, supported a speculation made 15 years earlier and dared to suggest that the scrapie agent contained protein but no genetic material, i.e. nucleic acids. The agent was given the name '**proteinaceous infectious particle**' or '**prion**' for short. This suggestion was unthinkable to many biologists and, in 1996, is still not accepted by a number of research workers who continue to search for an infective organism, such as a virus, which contains nucleic acids. However, the concept has been sufficiently accepted for the term 'prion disease' to be commonly used as a synonym for 'transmissible spongiform encephalopathy' disease (Section 5.1).

5.4.1 Properties of prion protein

The prion protein took some time to purify because it is insoluble and relatively resistant to degradation by proteases. Once identified, it was called '***p**rotease **r**esistant **p**rotein*' or **PrP** for short. The investigation of the role of PrP became more complicated when a protein with the same amino acid sequence but a different shape was found to be produced naturally. It seems that there is a single gene in mammals and birds which codes for the natural protein, here designated **PrP^cell^**, which is degraded by proteases. It is synthesized in all cells but particularly in brain cells where it is found attached to the external cell surface. In contrast, prion molecules, here designated **PrP^prion^**, collect together in deposits *within* cells which then no longer function normally.

The presence of PrP^{cell} in cells provides an explanation for how a protein agent devoid of nucleic acid can replicate itself (Figure 5.9). According to the hypothesis, PrP^{prion} in infected cells interacts with PrP^{cell} which then irreversibly changes its shape to form more PrP^{prion}. Thus, the PrP^{prion} has replicated itself in an unconventional way.

The normal role of PrP^{cell} in nature is unknown. Mice that have been genetically engineered to lack the PrP gene are perfectly healthy and normal in every respect, except that they are not susceptible to scrapie.

In inherited types of TSE diseases, mutations in the PrP gene result in forms of PrP^{cell} which spontaneously take the PrP^{prion} shape without intervention from an infective agent. In humans, familial CJD and Gertmann–Straussler syndrome (Table 5.1) are thought to be caused by genetic mutations of the PrP gene.

5.4.2 The species barrier

The **species barrier** is a term often used when discussing the susceptibility of new host species to prion diseases. It is of particular importance when considering the risk of transmission of scrapie from sheep to cattle and of BSE from cattle to humans.

The transmission of prion diseases *between* species appears to be unpredictable and is characterised by prolonged incubation times—as demonstrated by the sporadic incidence of the first cases of BSE.

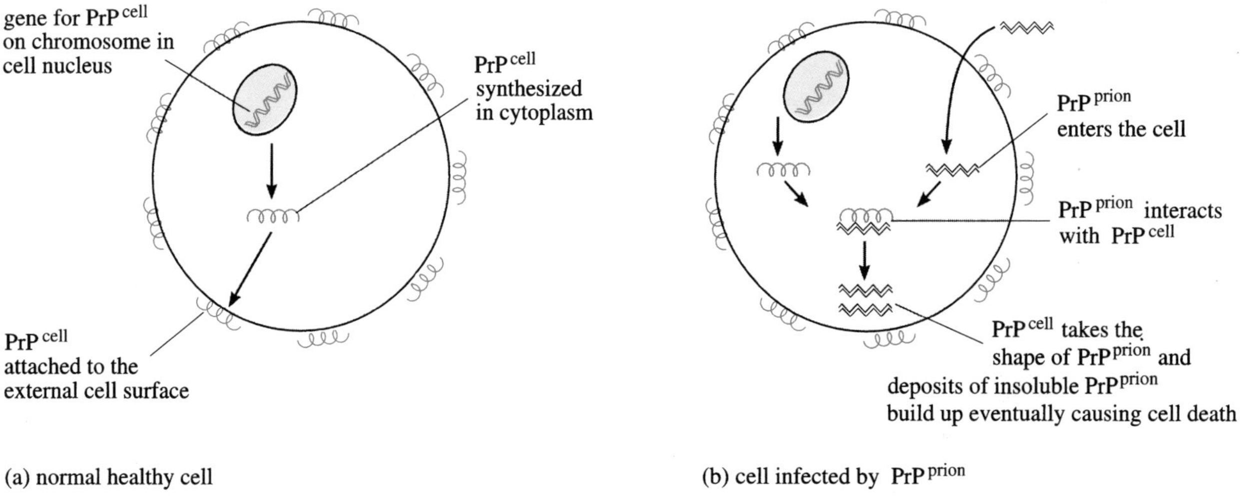

Figure 5.9 How PrP is thought to act within a cell. (a) The situation in a normal healthy cell. (b) The situation in a cell infected by PrP^{prion}.

▷ Why might this species barrier exist for prion diseases?

▶ The amino acid sequence in PrP^{cell} of a new host species may differ slightly from the amino acid sequence in the infective PrP^{prion}. When the host PrP^{cell} and the infective PrP^{prion} interact in the cell (Figure 5.9b), the two molecules may be incompatible (because of different amino acid sequences) and the conversion of PrP^{cell} to PrP^{prion} may be difficult.

On subsequent transmission of infected material within the new host species the incubation time often shortens. This can be explained because the infective PrP^{prion} from the *same* host species is compatible with PrP^{cell}.

▷ Which amino acid sequence will newly synthesized PrP^{prion} in the host cell reflect?

▶ The sequence in the host cell PrP^{cell}, as newly synthesized PrP^{prion} is derived from host cell PrP^{cell}.

This shortening of the incubation time can be reproduced experimentally:

(a) When hamsters (the new host species) were first challenged with PrP^{prion}-infected material from mice, the incubation time in the hamster was almost 400 days, i.e. transmission *between* species.

(b) When hamsters were subsequently challenged with PrP^{prion}-infected material from the hamsters in (a), the incubation time shortened to about 75 days, i.e. transmission *within* species.

Thus the species barrier seems to represent some incompatibility between the host PrP^{cell} and the infective PrP^{prion}.

5.4.3 Strain variation

An important argument against the prion hypothesis is the existence of different strains of scrapie. These maintain slightly different patterns of symptoms when passed through generations of a single strain of hamster and even between different species which have slightly different versions of PrP^{cell}. If, as we have seen, newly synthesized PrP^{prion} reflects the PrP^{cell} of the host, then the identity of the injected PrP^{prion} would be lost. Therefore, the question is 'how can the integrity of strains be maintained without the presence of nucleic acids?' Currently no one knows, despite ongoing research to address the issue.

For the BSE agent, only one strain has currently been described. It is different from all known strains of scrapie and it has maintained its identity through six different species.

5.5 The risk of transmission of BSE to humans

The government started to take control measures to reduce the risk to human health in 1988, two years after diagnosis of the first case of BSE, when regulations to destroy suspect animals and their milk came into effect (Table 5.3 and Table 5.5 overleaf). In November 1989, the Bovine Offal (Prohibition) Regulations prohibited the use of SBOs from cattle over six months of age in human food.

▷ When was the general public most at risk of consuming beef and beef products contaminated with BSE?

- Before November 1989; up to this point SBOs were still being added to human foodstuffs.

Table 5.5 Key dates and control measures introduced to protect human health.

Date	Control measure
Aug 1988	Carcasses of suspect animals destroyed.
Dec 1988	Milk from suspect animals destroyed.
Nov 1989	No SBOs to enter the human food-chain.
Apr 1995	Requirement to stain SBOs blue.
Aug 1995	The existing rules on SBOs consolidated and streamlined.
Nov 1995	Decision to prohibit the use of the backbone from cattle in the manufacture of mechanically recovered meat.
Apr 1996	Slaughter of cattle over 30 months of age in specially licensed plants and disposal outside the human and animal food-chains.
Jun 1996	UK government agrees to slaughter more cattle said to be at special risk of developing BSE.

It has been estimated, by some of those who believe that BSE is a cause for great public concern, that by the end of 1995 each member of the British public could, on average, have consumed over 80 meals containing beef from BSE-infected cattle. As the incubation period of BSE in cattle is three to six years and the majority of beef cattle are slaughtered at two-years-old or less, apparently healthy cattle incubating BSE could have been slaughtered and have entered the human food-chain.

- In which parts of cattle has infectivity been detected?
- In the brain, spinal cord and the retina from the eye but *not* in meat or milk (Section 5.3.3).

These infected tissues were most likely to have been included in meat products such as beefburgers, pies, sausages, soups, etc. before November 1989—after this date SBOs were excluded from the human food-chain.

However, in May and October 1995, during unannounced inspections of abattoirs, it was found that 48% of them were not complying with the requirements. Most were minor non-compliances, but in some premises violations were more serious; for example, SBOs were not separated and stored properly. Even though steps were taken to reinforce the regulations, this cause for concern led to questions in Parliament and further speculation by the media on the safety of beef. Other violations have also led to a further tightening of the regulations.

Following the announcement of the possible link of BSE with CJD in humans in March 1996 (Section 5.1), further measures were introduced to eliminate BSE in the United Kingdom and protect the consumer from any risk, however remote, of contracting CJD from beef from BSE-infected cattle. The measures are outlined in Section 5.3.3. The main regulation decreed that when cattle over 30 months of age are slaughtered, no parts of their carcasses can enter the human or animal food-chains.

There has been criticism of the way in which the government has handled the BSE epidemic. At the heart of this is the relationship between government ministers and their scientific advisers and this is examined in Activity 5.2.

Activity 5.2 *You should spend up to 20 minutes on this activity.*

Extract 5.1 presents the views of Richard Southwood, the government's chief adviser on BSE from 1988 to 1990, and details the way in which the Ministry of Agriculture, Fisheries and Food (MAFF) followed scientific advice and the consequences of this. Read the extract and answer the questions below. For some answers it may be useful to refer to Tables 5.3 and 5.5.

(a) Ministers insist that they have always done the scientists' bidding, but Southwood says that ministers were 'hostile' to many of his recommendations. Suggest how these views can be reconciled, using as examples the ban on SBOs from the human food-chain and the compensation given to farmers for the slaughter of animals with BSE.

(b) What evidence did Southwood use to conclude that the risk of BSE causing CJD in humans was remote and what was the consequence of this view?

(c) Did farmers cheat the system and pass off sick animals as healthy and, if so, why?

Extract 5.1 From *New Scientist*, 30 March 1996.

Ministers hostile to advice on BSE

Fred Pearce

A key government adviser on bovine spongiform encephalopathy (BSE) has dismissed claims by ministers that they have always followed the advice of scientists in their handling of 'mad cow disease'. This week an international outcry continued over the discovery that 10 Britons have died in recent months from a type of Creutzfeldt–Jakob disease (CJD) probably caused by eating beef contaminated with BSE. Some epidemiologists fear that thousands of people may now die from CJD.

Richard Southwood, the University of Oxford zoologist who was the government's chief adviser on BSE in 1988 and 1990, the years when diseased meat was passed on to humans in the greatest quantities, told *New Scientist*: 'I bridled when I heard ministers saying they were doing everything they'd been asked. The signals I got from senior civil servants at the time were very different.'

Ministers have defended their handling of BSE by insisting that they have always done the scientists' bidding. But Southwood says officials at the Ministry of Agriculture, Fisheries and Food (MAFF) delayed for two years before the committee he chaired to review the problem was set up in June 1988. And ministers were 'hostile' to many of his recommendations, Southwood says.

BSE was first diagnosed in cattle in 1986. At that time, following changes at the start of the decade in methods of manufacturing cattle feed containing animal protein, much of the national herd was eating feed containing scrapie. This is a long-standing disease among sheep, whose remains formed part of the cattle feed.

Scientists linked BSE to feed containing scrapie, but it was two years before the MAFF appointed Southwood to advise on its implications. It accepted his urgent initial recommendations to ban scrapie-contaminated feed and to slaughter infected cattle.

If BSE contamination has caused disease in people, most of the infection will have taken place in the late 1980s. Human exposure to BSE probably began to diminish only after November 1989, when the ministry tried to ban the inclusion of certain cattle organs, such as the brain and spinal cord, in food for human consumption.

The impact of the delays in government action has been compounded by BSE's long incubation period—typically four years—and possibly a longer period in humans. The result was a peak in cases of BSE in cattle in 1992, and a possible human epidemic that is only now beginning.

Southwood says it is not simply hindsight which suggests more could have been done sooner to halt the hidden epidemic. Regretting the government's two-year delay in appointing his committee, he says, 'I think if we had been asked in 1986 we would have recommended taking diseased carcasses out of the food cycle at that stage.'

Southwood has been criticised for not recommending from the start that there should be a ban on cattle brains in meat products for human consumption. But he says that such recommendations would have stood little chance. 'We felt it was a no-goer. They [MAFF] already thought our proposals were pretty revolutionary.' He later pressed for the ban, which was finally approved in November 1989. 'The ministry wanted a one-year ban,' he says, 'I had to insist that it was permanent'.

Southwood also advised ministers that farmers should be given full compensation for infected cattle that were destroyed. But he says MAFF 'would not do it'. Instead it offered a 50 per cent grant. The Earl of Lindsay, a Scottish Office minister, admitted last week that this cash-saving policy encouraged farmers to pass off sick animals as healthy, and prolonged the period during which humans were fed potentially lethal beef.

Southwood's key conclusions in 1989 remained in place until last week. In 1989 he said that the possibility of BSE causing disease in humans appeared 'remote', because humans had been exposed to scrapie for many decades without harm. 'The scrapie model looked the most likely one,' he now says, 'but it looks as if we may have been wrong.'

But he says it was the small chance of humans succumbing that drove his recommendations, especially that the potentially infected parts of cattle should not be fed to humans and that there should be a programme to monitor CJD. The CJD Surveillance Unit in Edinburgh was set up in 1990 to 'evaluate changes in patterns' in the occurrence of CJD 'which might be attributable to BSE'. It was Rob Will, the head of that monitoring programme, who blew the whistle earlier this month after finding 10 unusual cases of CJD.

However, the conclusion that the risks to humans were remote led to complacency through the early 1990s. Farmers carried on using contaminated feed for many months, and evaded bans on selling diseased animals.

Abattoirs bent the rules intended to keep the potentially lethal parts of cattle separate from material destined for human consumption. And officials were less than diligent in policing their rules, and put off closing potential loopholes for the disease.

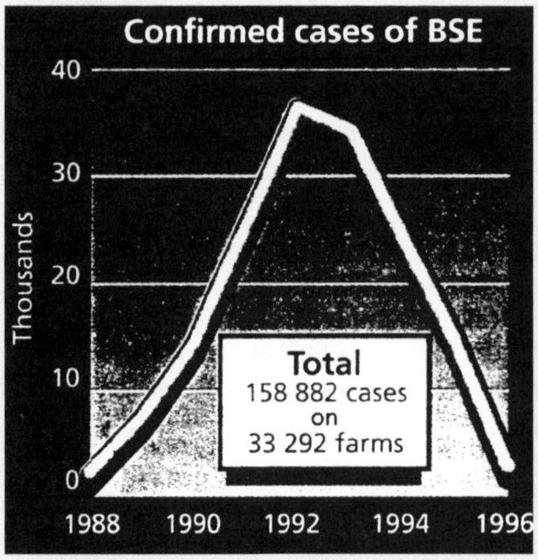

Some microbiologists were this week claiming that complacency also led the CJD monitoring unit to ignore evidence that people were dying from unusual strains of CJD as long as five years ago. Worst of all for the reputation of the government today, it led ministers to proclaim that beef was safe.

The scientists on the Spongiform Encephalopathy Advisory Committee, set up in 1990 as a permanent advisory group, insist that they have not changed their tune about human risks. Jeffrey Almond of the University of Reading wrote last November that 'the possibility that BSE might transmit to humans has been acknowledged since the disease was first recognised in British cattle'. He said that scrapie, BSE and related diseases in other animals 'can transmit from one species to another' but that 'for any given donor species there is no way of predicting which recipient species will be susceptible'.

The government's chief medical officer Kenneth Calman agreed last week that 'the theoretical risk has always been acknowledged'. But his stock response to questions has for several years been that 'there is no evidence that BSE can cause CJD in humans'.

Down the long whispering corridors of Whitehall, a message that began 'there is no way of predicting ...' seems to have reached the other end as 'there is no evidence ...', which is a subtly different proposition. It is an error that may spell disaster for the reputations of ministers and the future of the British beef industry—not to mention the lives of possibly thousands of people who were told that British beef was safe to eat.

5.5.1 Research in progress

It is, of course, impossible to settle the issue of the risk of transmission of BSE to humans by carrying out experiments on humans themselves. As they are related species, one approach is to study the effect of challenging other primates. In recent experiments, macaque monkeys, which had been infected with brain extracts from BSE-infected cattle, showed a very similar brain cell pathology, i.e. the distribution of spongiform changes, to the new variant of CJD first diagnosed in Britain during 1994 and 1995. This is further evidence of a possible link between BSE and CJD in humans.

An alternative approach is to genetically engineer mice to contain the human gene for PrP. This method does not, of course, give a full human representation but does allow investigation of the species barrier. Mice also have a relatively short incubation time for BSE. In reports of ongoing experiments, both genetically-engineered mice carrying the human gene for PrP and wild-type mice have survived for more than 450 days after injection of BSE into their brains. However, as wild-type mice can survive for 600–700 days before demonstrating clinical symptoms, it will be later in 1996 before the final experimental results become available. Then we may have a clearer understanding of the level of risk to humans from consumption of beef.

Summary of Chapter 5

1 The new disease, BSE, was first diagnosed in cattle in early 1985. It belongs to a group of TSE diseases of mammals which affect the central nervous system. These include CJD in humans and scrapie in sheep.

2 In March 1996, a possible link was established between BSE in cattle and CJD in humans, following the diagnosis, in 1994 and 1995, of a new variant of CJD.

3 Because of this possible link, consumers lost confidence in the safety of beef and for a time stopped buying it. The EU imposed a world-wide ban on the export of British beef and beef products.

4 Early epidemiological studies showed that exposure started in winter 1981/82 with no distinct geographic focus. Dairy calves had the greatest level of exposure and incubation times were long (2.5–8 years).

5 Investigation into the possible cause led to the formulation of the hypothesis that 'BSE came from consumption of concentrated feed by dairy calves which probably contained scrapie agents in MBM derived from infected sheep'.

6 From 1980, following changes in the regulations controlling the rendering process of waste materials from abattoirs, the hydrocarbon solvent extraction process was gradually phased out. The resulting single process is thought to be less effective at deactivating the scrapie agent.

7 In bioassays of tissues from infected cattle, infectivity has been detected in the brain, the spinal cord and the retina.

8 Measures were introduced to control animal health. Key ones were the ban on the use of ruminant protein in ruminant feed (July 1988) and payment of 100% compensation for the slaughter of animals suspected of having BSE (February 1990). It was thought that with these measures the disease would disappear.

9 However, cattle born after the ban continued to develop BSE. This has been attributed to the use of contaminated feed after the ban and a disregard of regulations for dealing with SBOs in abattoirs.

10 The existence of BSE in cattle born after the ban could also be explained if there was direct transmission from animal to animal in a herd and/or … or from parent to offspring. Indirect evidence suggests that if direct transmission does occur it is at a low level.

11 The BSE agent, or prion, is thought to be an aberrant form of a natural protein, PrP, which is produced by mammals and birds. It is resistant to degradation by proteases and replicates itself by changing the form of the natural protein to that of its own.

12 A prion from one species can, under some circumstances, break the species barrier and infect another species. Because of an incompatibility of the prion PrP (from one species) with the host PrP (from another species), this process is unpredictable and is characterised by long incubation times. Once the species barrier has been broken and the prion PrP is from the same species as the host, infection is predictable and incubation times are shorter.

13 In November 1989, SBOs were banned from inclusion in the human food-chain in order to protect humans from the risk of contracting CJD from products of BSE-infected animals. Following the announcement of a possible link between BSE and CJD in March 1996, carcasses of all cattle over 30 months old were banned from the human food-chain.

14 The greatest danger to humans of contracting CJD from contaminated meat came from eating beef before November 1989, but because of the long incubation time of TSE diseases the true risk is still unknown.

15 Current research into the transmission of BSE to humans uses monkeys, which are closely related to humans, and mice genetically engineered with the human PrP gene as models.

6 Pesticides (5 hrs)

6.1 Introduction

Having been 'informed' mainly by television programmes and (to a lesser extent) by newspapers and magazine articles, it seems that many people are worried about the use of **pesticides**, whether in agriculture, public health, industry or around the home. It is, however, important to realize that such sources of information may not be free from bias. Raising alarm undoubtedly makes more compelling viewing and more dramatic headlines than do more dispassionate treatments.

Innovations always entail risk. A balance therefore has to be struck between the level of risk associated with a change and the likely benefits accruing from it (a process sometimes called **risk/benefit analysis**). Few people would seriously propose banning the use of electricity, the motor car or blood transfusion (to take just three innovations of the past). It is, however, entirely reasonable to insist that the inherent dangers of any innovation be reduced as far as possible and certainly to an acceptable level. Since the problems associated with the use of some of the first synthetic pesticides became apparent, much has been done to replace them with compounds that are less dangerous to humans and less environmentally hazardous, to raise safety standards of application and to develop alternative methods of controlling agricultural problems to enable a reduction in pesticide use. There have also been rapid developments in the techniques of monitoring and forecasting changes in pest populations in order to reduce unnecessary spraying.

In this brief discussion, we will follow agricultural pesticides through from their discovery and development to their fate in the environment. The chapter concentrates on **insecticides**, but also considers **fungicides** and **herbicides** (i.e. weedkillers). It is first necessary, however, to consider *why* pesticides are needed at all.

Any human-induced change to the natural world—and all forms of agriculture are just that—significantly alters the habitat, so that many of the original species of animals and plants can no longer survive and are replaced by a different, less stable, community. In contrast to mixed natural vegetation, a growing crop provides an abundance of one type (a **monoculture**) which can be exploited by other organisms to such an extent that ecological imbalance can occur. This can result in a lowering of the quality and/or the yield of the crop, and hence in the exploiting organisms becoming regarded as pests. It is the consumer's demand for high quality at low prices which makes the use of pesticides to control these organisms necessary.

If pesticide use were stopped, the price of agricultural commodities would rise significantly and their quality would deteriorate. However, by ensuring that pesticides are used in ways which favour their action and minimize their impact on the environment, it is possible to reduce their use and still achieve control of pest species without undue hazard to other organisms or to the environment in general.

6.2 The discovery and development of pesticides

The pesticides used at the turn of the century were a mixed bag of inorganic compounds (e.g. sulphur, arsenates) or toxic extracts of plants (e.g. nicotine). Many of the insecticides derived from knowledge of their toxicity to humans and other animals (e.g. arsenates, plant extracts used to tip hunting arrows). They were subject to very little safety testing. Fortunately, the plant extracts were rapidly broken down in the environment after application, and anyway the use of pesticides at this time was relatively limited.

The big change came in the late 1930s with the exploitation of the insecticidal properties of DDT (a molecule first synthesized in 1874). Compared to the very short life of the plant extracts previously used, the long-lasting effects of DDT appeared a virtue. Farmers quickly took to this cheap and persistent chemical which provided effective pest control. The way was open for the development of an industry to supply a complete range of chemical solutions to agricultural problems.

Today, many hundreds of compounds are tested each year by the companies involved in developing new pesticides. Some of these compounds are developed by the pharmaceutical industry, some are by-products of other chemical processes and some are new molecules synthesized as part of university research. They are usually tested simultaneously in different laboratories in a fairly simple initial screening for any insecticidal, fungicidal or herbicidal activity against a range of potential target types (e.g. sucking and chewing insects, fungal diseases such as rusts and mildews, grasses and broad-leaved plants). Most show no activity, or turn out on further investigation to have been through the test on a previous occasion. Very few are taken to the next stage, where they are tested in considerable detail to establish whether they have a potential market and are adequately safe for humans and for the environment. Here they are joined by other compounds 'engineered' on the basis of the structure of previously successful groups of chemicals.

Testing past the initial stage becomes increasingly costly, especially tests for human toxicity and large-scale international field trials; at the time of writing, an average figure for the whole process from discovery to marketing is about £32 million per marketed pesticide. This is still an economically viable process for herbicides and frequently so for fungicides, but it is doubtful how often it proves profitable for insecticides. New insecticides may lose their initial clear superiority over cheaper, older compounds quite rapidly as levels of resistance to the chemical develop in the pests.

Following the testing process, a portfolio of data is submitted to government registration authorities in order to obtain permission for the product to be marketed. The requirement by different countries (or even different states within a country, e.g. Australia) for different information inevitably adds to the costs. Much of the testing is, of course, concerned with efficacy against the target organism, coupled with checks that there is no obvious damage to the crops concerned. Stringent short- and long-term tests on laboratory animals are used to identify the doses which might cause acute or chronic illness or death to humans; special emphasis is placed on testing for carcinogens and for effects on the foetus in pregnant animals. The results of such tests provide the safety data for handling and applying the compound. Very extensive data are also collected on the breakdown of the pesticide, both in plants and in the soil; the plant results are essential for determining the safety to the consumer of the crop at harvest, while those in the soil are used to assess the degree of potential hazard to the environment. This research is complicated by the fact that the breakdown products of some pesticides may actually be *more* toxic than the parent compound; in the end, for any one pesticide under investigation, there may be quite a number of compounds whose fate in plants and in the environment will need to be traced.

Extensive research is also conducted on the environmental impact of new pesticides. For example, the effects on fish, birds and other wildlife, beneficial insects (e.g. bees) and soil microbes have to be assessed. Special efforts are made to detect any evidence of pesticide accumulation in food chains. Much of this is quite long-term work and it continues well after the pesticide in question has been marketed.

Manufacturers of pesticide products are strongly motivated towards providing formulations which show maximum advantages and have minimum side-effects. Governments throughout the world require pesticides to be registered, so that both their efficacy and safety can be monitored. There is internationally accepted legislation covering the safety of the operator applying the treatment, the conditions under which the treatment may be applied, and the precautions to be taken to prevent contamination of food or the general environment. After consideration of all the relevant effects on non-target organisms arising from appropriate use of the product, **maximum residue levels** defining the concentrations allowed in and around food have been set for every registered pesticide. Thus, as long as a pesticide is used according to its label instructions, these considerations provide a safeguard against the occurrence of any reasonably predictable (as well as most unforeseen) adverse effects arising from its use. All data concerning methods and rates of application, safety, etc. are summarized on the product label which has been subject to approval by registration authorities. *It is therefore essential that the label is read carefully before any pesticide is used.* If unforeseen effects are noted after any pesticide product is marketed, then the registration is immediately subject to re-assessment.

Question 6.1 Newspapers sometimes report that a pesticide has been withdrawn after several years of use because of an unforeseen hazard. What do you think our reaction should be to such news?

6.3 The contents of the poison cupboard

This section provides a brief overview of the insecticides (6.3.1), fungicides (6.3.2) and herbicides (6.3.3) principally in use today. You should concern yourself mainly with the *major chemical groups of pesticides.* You should certainly *not* attempt to learn either the *names* of individual pesticides (which are given only to provide examples of the groups) or their *structures* (which are included only to give you a feel for the relative complexity and relative diversity of synthetic pesticides). Details such as the chemical structures, properties, formulations, uses and toxicology of all current pesticides can readily be found in *The Pesticide Manual* (Worthing, C. R. and Hance, R. J. (1990) 9th edn, British Crop Protection Council, Farnham).

It should be noted that pesticides are also available to control less important pests (e.g. eelworms, molluscs, birds) and that insects (for example) are sometimes controlled using quite novel substances (e.g. insect disease agents, analogues of insect hormones).

6.3.1 Insecticides

Some of the earliest insecticides were **natural plant extracts** such as *pyrethrum* (from a species of *Chrysanthemum*), *rotenone* (from a species of *Derris*; hence 'derris dust') and *nicotine* (from a species of *Nicotiana*, tobacco). Such extracts are not necessarily less toxic to humans than industrially synthesized insecticides; however, they are broken down quite rapidly and so are of short **persistence** in the plants to which they are applied. The use of these insecticides is currently increasing with the growth of organic agriculture and horticulture. There are also situations in which they are still used by non-organic growers.

▷ When might a non-organic grower need to use one of these short-lived insecticides rather than a more persistent synthetic alternative?

▶ When it is necessary to apply an insecticide near to the time the crop is harvested.

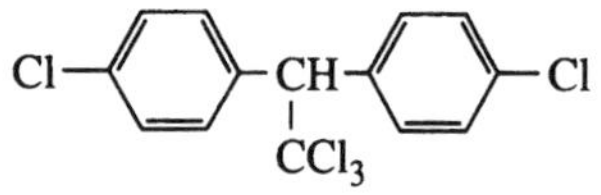

DDT, an organochlorine

The first synthetic insecticides were the **organochlorines**, of which *DDT* (dichlorodiphenyltrichloroethane) is the most famous—or infamous, depending on your point of view. DDT is cheap to manufacture and relatively safe to handle. Organochlorines are not readily degraded in the environment and can therefore give long-lasting protection. However, their very persistence, which was hailed as a major breakthrough when they were first introduced, also led to their downfall. Although relatively non-toxic to fish, birds and mammals when taken directly, they tend to build up in concentration along food chains (i.e. concentrations are higher in insectivorous birds and mammals than in insects, and higher still in animals which prey on the insectivorous species—a process sometimes referred to as **biomagnification**). The effects on birds of prey were often sub-lethal; i.e. although they were not killed, they were adversely affected in other ways, most commonly through reduced breeding success (probably because of more broken eggs as a consequence of eggshell thinning). It was this sensitivity of birds to organochlorine insecticides which highlighted the problem and caused their withdrawal in many countries from the 1960s onwards. The insecticides were, in any case, becoming less useful because many pests had become resistant to them. It was also realized that some (e.g. DDT) were just as lethal to the pests' natural enemies. *Dieldrin*, which has a persistence of many decades, was banned quite early on. It was later found that the more rapid disappearance of a related compound, *aldrin*, was partly due to its conversion into dieldrin within the soil. The persistence of the organochlorines would still be an asset for certain insect pest problems, particularly in the soil (and on bulbs) where treatment often has to precede pest appearance by a considerable time. *Endosulfan*, which has the desirable attribute of doing relatively little damage to parasites of crop pests, is still widely used as a general insecticide in the tropics.

Question 6.2 DDT is still exported from Europe to the tropics. How do you feel about the fact that we export a pesticide that we are not prepared to use ourselves?

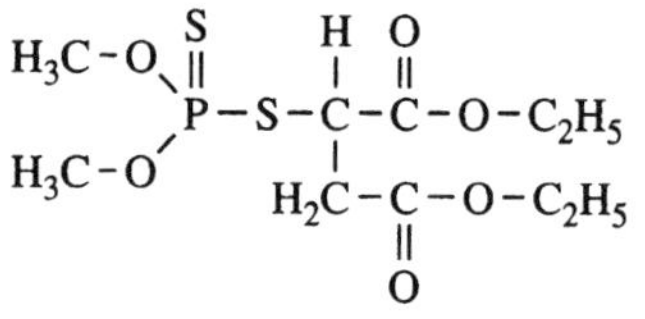

malathion, an organophosphate

The **organophosphates** were the next group of synthetic insecticides to be developed. Although the first organophosphates were very toxic to humans, less toxic members such as *malathion* have since been developed and are widely used. Nevertheless, they do need handling with care and protective clothing is usually necessary. One of the advantages of this group is that it offers a choice of *routes to target* (see below). Organophosphates are much less persistent than the organochlorines, although some can be incorporated into granules which dissolve slowly in the soil to provide long-lasting protection.

Question 6.3 Why is it more 'environmentally friendly' to use granules of an organophosphate for long-lasting protection than to incorporate an organochlorine pesticide into the soil?

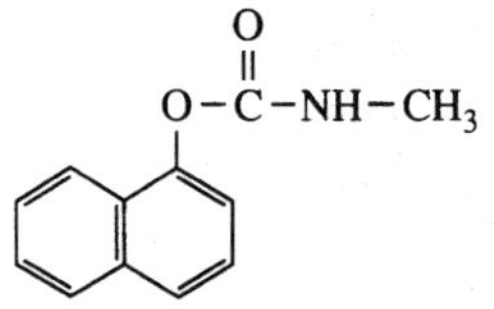

carbaryl, a carbamate

The **carbamates**, which are intermediate between the organochlorines and organophosphates in terms of both human toxicity and persistence, came next with the introduction of *carbaryl*. Relatively few carbamate compounds have proved commercial; while some are effective *systemic* insecticides (see below), they are dangerously toxic to humans.

The most recent group of synthetic insecticides are the **synthetic pyrethroids** such as *cypermethrin*. These are based on the structure of natural pyrethrum, but are more stable to sunlight. They are very toxic to insects and can be applied to crops in very small quantities compared with the earlier insecticides. Their persistence is only a matter of days or a few weeks, and they are relatively safe to humans. However, a severe limitation is that only one route to target is available using them (the *residual*;

see below). Also, their high toxicity to insects has resulted in the same kind of loss of natural enemies experienced with DDT and, where they are used routinely, in rapid development of resistance.

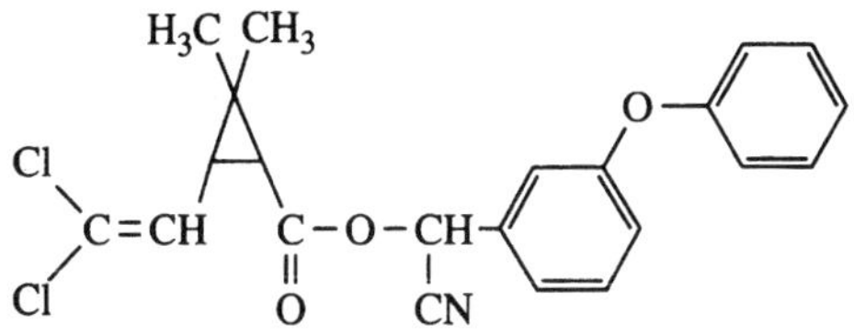

cypermethrin, a synthetic pyrethroid

▷ All of these groups of insecticide kill by paralysing the nervous system. What is the advantage of nerve poisons compared with insecticides that might operate in different ways?

▶ Paralysing the nervous system gives a very rapid kill, thereby dealing with the problem caused by the pest within hours rather than days or weeks.

Since most insecticides will kill any insect fairly rapidly in the laboratory, the choice of which to use is influenced more by other factors such as **route to target**. The main routes are listed below.

Contact The insecticide penetrates the cuticle of the insect on direct contact.

Residual The insecticide remains on the leaf for a while and can be picked up by an insect walking over the residue.

Translaminar The insecticide passes through the leaf from the upper surface (to which it is more easily applied) to the lower surface (where many insect pests reside).

Systemic The insecticide is absorbed by the plant and then moved around (i.e. **translocated**) in its own transport systems. In fact, nearly all systemic insecticides are transported in the **xylem** system that carries water and minerals from the roots to the rest of the plant rather than in the **phloem** system that carries organic nutrients from where they are produced or stored (e.g. leaves, tubers) to where they are utilized (e.g. roots, developing fruit).

Fumigant Vapours of the insecticide are 'inhaled' by the insect.

Activity 6.1 *You should spend up to 20 minutes on this activity.*

Table 6.1 summarizes the results of an experiment designed to reveal the route(s) by which a particular insecticide reaches its target—in this case, the black bean aphid. What route(s) to target does this insecticide take?

In fact, interpreting a table as complex as Table 6.1 is not a trivial task and to be able to do so confidently requires practice. It is often best to analyse the data in fairly small steps.

(i) What is meant by the term 'replicate plants A–F' and what purpose is served by replication?

(ii) What purpose is served by treatment 1? What general term is applied to such treatments in an experiment?

(iii) Match each of treatments 2 to 8 with one of the routes to target listed above.

(iv) For each of treatments 1 to 8, calculate the mortality of the aphids as a percentage (don't forget to take *all* the replicates into account) and complete the final column of Table 6.1.

(v) By comparing the routes to target (represented by the various treatments) with the percentage mortalities, you should be able to answer the original question: 'What route(s) to target does this insecticide take?'

Table 6.1 Results of an experiment in which 10 black bean aphids (*Aphis fabae*) were caged on each of 6 replicate plants (A–F) of broad bean (*Vicia faba*) for each of 8 different treatments (1–8) with a particular insecticide.

Treatment	Number of dead aphids (out of 10) on each replicate plant 48 hours after treatment						Mortality /%
	A	B	C	D	E	F	
1 Aphids caged on underside of leaf of untreated plant, i.e. no insecticide used.	2	2	2	3	0	2	
2 Insecticide poured onto soil at base of plant on which aphids caged on underside of leaf.	10	10	10	10	10	10	
3 Aphids sprayed, allowed to dry and then caged on underside of leaf of untreated plant.	10	10	10	10	10	10	
4 Insecticide painted on underside of leaf; aphids caged on underside of leaf when insecticide dry.	10	10	10	10	10	10	
5 Insecticide painted on top side of leaf with aphids caged on underside.	8	9	7	5	10	7	
6 Insecticide painted on underside of leaf at tip; aphids caged on underside of leaf at base.	2	5	3	2	2	0	
7 Insecticide painted on underside of leaf at base; aphids caged on underside of leaf at tip.	10	10	10	5	9	7	
8 Aphids caged on underside of leaf of plant inside cylinder with insecticide-soaked filter paper as lid.	10	10	10	10	8	9	

6.3.2 Fungicides

Fungicides are conveniently split into **protectants** (i.e. effective at the site of application) and **systemics** (i.e. translocated around the plant). The protectants are all general toxins that usually cause extensive cytoplasmic damage to the fungi; any selectivity results from selective uptake and accumulation. They often affect fungal spores and therefore protect against attack rather than eliminate established infections. The systemics are often curative as well as protective. However, they are also more biochemically selective and, as a result, several problems of resistance to systemic fungicides have arisen.

▷ Why are resistance problems more likely with relatively selective systemic fungicides than with relatively non-selective protectants?

▶ It is much more likely that a fungus will develop, through natural selection, some mechanism for protecting a specific biochemical site than for resisting extensive cytoplasmic damage. One reason for this is that the latter would probably require changes involving many more genes than would the former.

Protectants

Metal- or **sulphur-based fungicides** were the first protectants to be developed. Copper-based *Bordeaux mixture* has been used for over a century and *sulphur* has also been used very extensively. (The fact that such inorganic compounds as these may be used in *organic* farming illustrates the confusion surrounding the term 'organic', which in this context is a concept unrelated to the words 'organic' and 'inorganic' as used in chemistry.) Organic chemicals incorporating tin or mercury were developed as fungicides somewhat later.

A second group of protectants are the **phthalimides** (e.g. *captan*, which was developed in the early 1950s and is used extensively against apple scab) and the **dicarboximides** (e.g. *iprodione*).

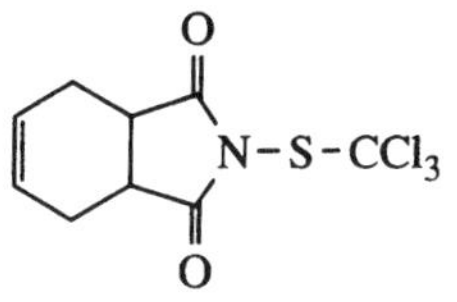

captan, a phthalimide

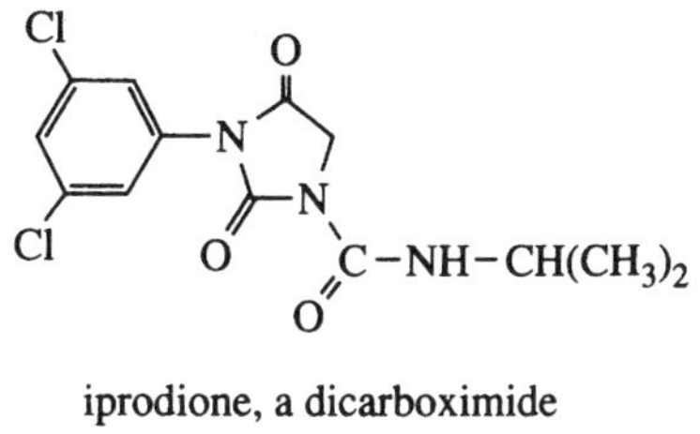

iprodione, a dicarboximide

Systemics

The first systemic fungicides to be synthesized were the **oxathiins** (e.g. *carboxin*), which are specific against smut diseases, and the **hydroxypyrimidines** (e.g. *ethirimol*), which are specific against powdery mildews. **MBC generators** (e.g. *benomyl*) are a group of fungicides which specifically inhibit microtubule assembly in some groups of fungi. They are called MBC generators because they are broken down in the fungus to the active fungicidal agent, *m*ethyl 2-*b*enzimidazole*c*arbamide. **Ergosterol inhibitors** (e.g. *triadimefon*) are a large, chemically diverse, group of fungicides that inhibit the synthesis of ergosterol, an essential component of the membranes of several groups of fungi. **Phenylamides** (e.g. *metalaxyl*) have been considerably exploited as systemic fungicides for control of potato blight and downy mildews.

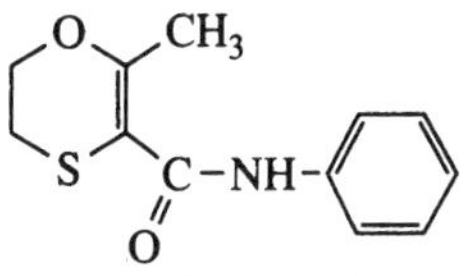

carboxin, an oxathiin

Nowadays, many systemic fungicides (particularly the phenylamides) are sold as mixtures with protectants, following earlier problems of resistance building up in the targets—presumably because fewer organisms survive to develop resistance to the systemics.

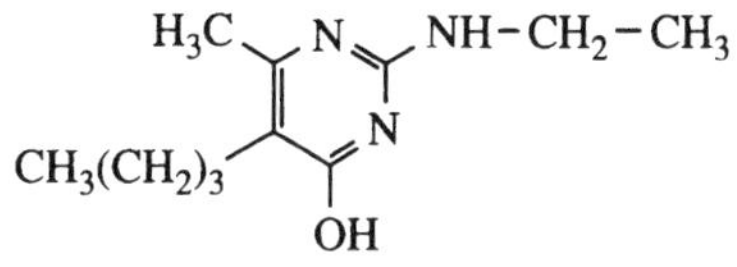

ethirimol, a hydroxypyrimidine

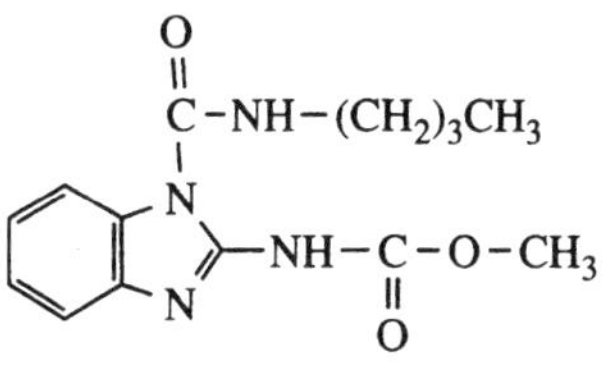

benomyl, an MBC generator

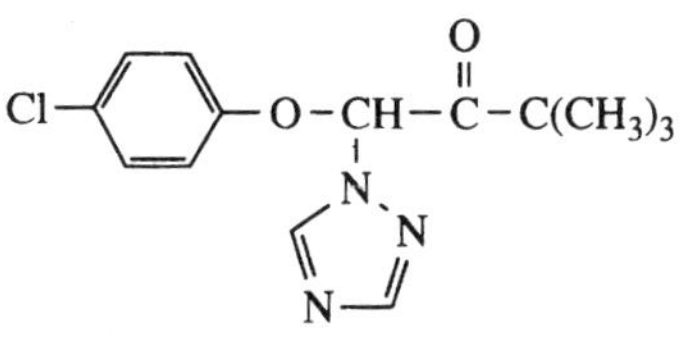

triadimefon, an ergosterol inhibitor

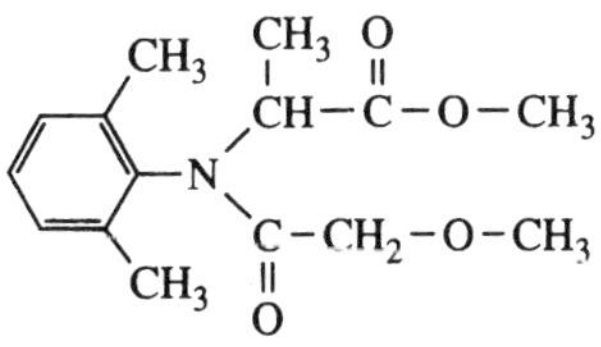

metalaxyl, a phenylamide

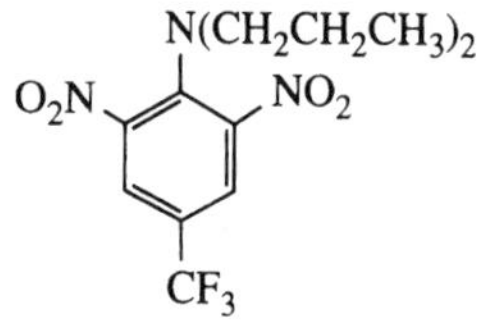

trifluralin, a dinitroaniline

6.3.3 Herbicides

Chemicals with a general toxicity to plants can be used as total weedkillers in order to clear weedy ground—either to remove weeds before the crop emerges (i.e. pre-emergence) or sometimes to kill weeds between woody stems (e.g. in forestry or fruit growing). Volatile compounds, such as the **dinitroanilines** (e.g. *trifluralin*), which quickly evaporate from the soil, are required for pre-emergence treatment. Less volatile compounds, such as the **triazines** (e.g. *simazine*) and **ureas** (e.g. *diuron*), are used to keep the soil clear of vegetation for most of a season.

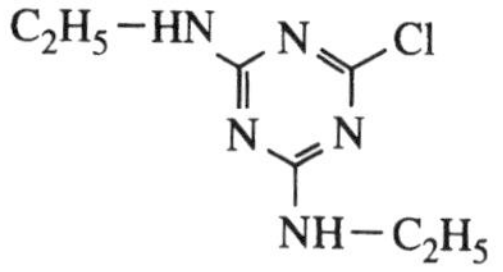

simazine, a triazine

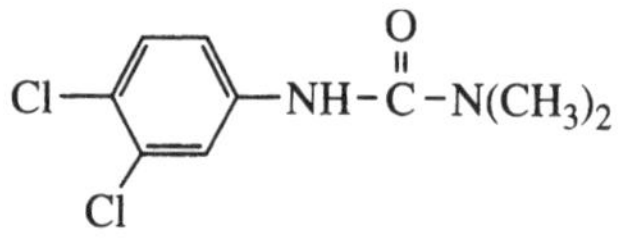

diuron, a urea

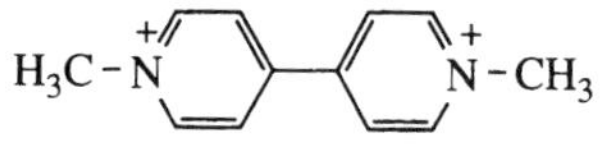

paraquat, a bipyridinium herbicide

The **bipyridinium herbicides,** such as *paraquat* and *diquat*, interfere with photosynthesis and cause very rapid plant death. (To all intents and purposes, they act as 'chemical flame-guns'.) However, these herbicides are effectively inactivated by becoming strongly bound to clay minerals in the soil; therefore they do not affect emerging seedlings (or transplanted crops)—but unfortunately, neither do they prevent the regrowth of perennial weeds from tap roots.

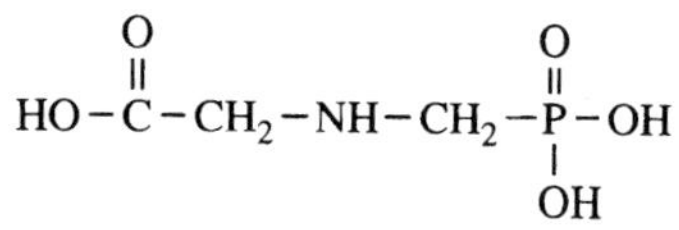

glyphosate, an organophosphate

Glyphosate is an **organophosphate** which is similarly inactivated on contact with the soil, but it is an unusual systemic pesticide in that it is translocated *downwards* in plants. Although it is much slower acting than the bipyridinium compounds, it does have the advantage of killing the roots of perennial weeds.

Often, however, a herbicide is needed to deal with weeds *in* a crop. It is quite difficult to find compounds which select between species. Nevertheless, by judicious choice of compound and application method, it is possible nowadays to select against wild oat (*Avena* spp.) as a weed in cereal crops, even though all are closely related members of the grass family (the Gramineae).

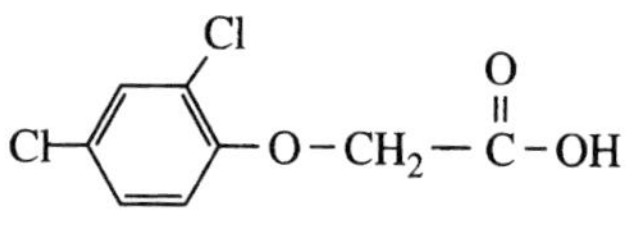

2,4-D, a phenoxyacetic acid

Since the **phenoxyacetic acids** (e.g. 2,4-dichlorophenoxyacetic acid, *2,4-D*) are selective against **dicotyledonous plants** (broad-leaved plants which have two seed-leaves) they have been used widely for weed control in cereals (which are **monocotyledonous plants**, having narrow leaves and one seed-leaf). These herbicides mimic natural plant growth regulators but produce weird twisted growth. A frequently quoted explanation for their selectivity is increased herbicide retention on the broad leaves of the weeds compared to the narrow leaves of the cereals, but the story is actually more complicated than this.

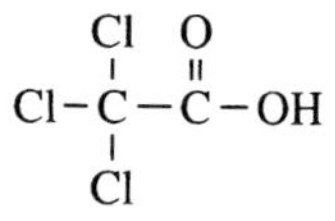

TCA, a chlorinated aliphatic acid

It is also possible to control grass weeds in a broad-leaved crop by using **chlorinated aliphatic acids** (e.g. trichloroacetic acid, *TCA*, and dichloropropionic acid, *dalapon*). However, the basis of this selectivity is not clear.

6.4 Application of pesticides

The active ingredients of pesticides, as described in Section 6.3, are never applied to plants 'neat'. Most are intended to be dissolved in an inert 'carrier' (often water, but sometimes oil) before use. In any case, the active ingredients are always **formulated** with additives such as detergent-like compounds to wet plant leaves (which helps spread the pesticidal film) and emulsifiers (which prevent oil-soluble pesticides separating rapidly when mixed with water). Very insoluble compounds may be sprayed in water either as finely milled particles or as coarser 'wettable' powders. Finally, some pesticides are applied in a dry form as dusts, granules or seed dressings. The formulation and method of application can profoundly influence pesticidal action. Here we will restrict our attention to spraying, which is the commonest form of pesticide application.

It comes as a surprise to most people to discover how little of a spray actually lands on plants. It depends a lot on the plant species, its stage of growth and the weather conditions, but a figure of 1.5% is probably optimistic in many situations. To put this more dramatically: for every £100 farmers spend on pesticide, £98.50 is wasted.

Is this inevitable? Shouldn't something be done about it? The answer to the first question is a qualified 'yes' and the answer to the second (see below) is 'something *can* be done about it, but it may not be simple enough to become widely adopted'.

The basic problem is that conventional spraying involves forcing a liquid under pressure through a small hole in order to convert *one large 'drop'* (the contents of the spray tank) into *many smaller drops*. The smaller the drops, the greater the number that can be produced from the spray tank, and therefore the better the coverage of the field to be sprayed.

Activity 6.2 *You should spend up to 15 minutes on this activity.*

(a) If it were possible to convert one litre of spray into identical 1 mm diameter spherical drops and then spread these evenly over an area of one hectare (100 m × 100 m), how many drops would there be on each 1 m^2?

If you cannot immediately see how to answer this question, then try tackling it as a series of smaller questions:

(i) What is the volume (in mm^3) of each of the identical 1 mm diameter spherical drops? In case you cannot recall it, the formula for the volume (V) of a sphere is $V = \frac{4}{3}\pi r^3$, where r is the radius.

(ii) What is 1 litre expressed in mm^3?

(iii) How many identical 1 mm diameter drops can be obtained from 1 litre of spray?

(iv) How many 1 m^2 are there in 1 hectare?

(v) How many identical 1 mm diameter drops would there be on each 1 m^2?

(b) Suppose the drops were 100 times smaller in diameter (i.e. their diameters were 0.01 mm). How many drops would there be on each 1 m^2?

The physics of forcing liquid through a small hole inevitably produces drops of a wide range of sizes. There are myriads of tiny drops, but the few large ones can account for much of the volume of liquid being sprayed (as you saw in Activity 6.2, a drop with a diameter 100 times greater than another would have a volume a million times greater). More than half the *number of drops* from a typical spray nozzle might have diameters less than 50 μm (i.e. 50×10^{-6} m which is 50×10^{-3} mm or 0.05 mm), yet more than half the *volume* being sprayed could be in the form of the relatively few drops with diameters greater than 200 μm.

The small drops have little mass and quickly lose the momentum imparted from the spraying machine. They therefore get carried away by any air movement, which in daytime is usually upwards. Since drops begin to evaporate and lose mass as soon as they leave the nozzle, this upward drift may apply to drops as large as 80 μm in diameter at the point of production. Moreover, even drops large enough to reach plants may not land or be retained. Smaller drops stream around obstacles to air-flow (e.g. leaves and stems). Larger drops (possibly from 200 μm diameter upwards) bounce off any leaves they touch (indeed, it is possible to apply weedkiller to the soil of a dense crop by allowing 400 μm diameter drops to bounce their way from leaf to leaf and eventually onto the soil). Furthermore, if a leaf becomes so fully wetted that the drops flow into each other (i.e. coalesce), most of the liquid will run off to the ground leaving just a very thin liquid film on the leaf.

The result of all this is that, of the entire spectrum of drop sizes produced by a spray, only those around 100–120 μm in diameter are likely to be deposited on plants effectively. The inefficiency referred to earlier is accounted for by so much of the volume of the spray being in drops outside this size range.

Question 6.4 To allow for the losses inherent in traditional spraying, application of around 750 g of active ingredient per hectare is recommended for many pesticides. Assuming the earlier (optimistic) estimate for the percentage of spray landing on the plants, what is the effective amount (in $g\,ha^{-1}$) applied to the crop?

Theoretically, much of the resulting waste could be avoided if we gave up forcing liquid through holes and found a different way of generating drops, more of which are the right size. There are at least three ways of approaching this ideal.

6.4.1 The spinning cage

The first way is to chop a flow of liquid into pieces by allowing it to pass through a rapidly rotating cage made up of many knife-edges. This is the principle of the 'spinning cage' nozzle (Figure 6.1), originally developed for spraying from aircraft. The cage is made of fine-mesh metal gauze and is spun by propeller blades rotated by the air-flow over the wing of the aeroplane. The device can, however, be employed on ground equipment by spinning the cage with a direct drive. The spinning cage produces what is still a fairly wide size range of drops, but there is a considerable reduction in the number of both very tiny and very large drops.

Figure 6.1 The spinning cage.

6.4.2 The spinning cup

Another approach is to use centrifugal force to shear fine 'filaments' of liquid into drops. This approach is typified by the 'spinning cup' (Figure 6.2), in which liquid dribbles into a grooved cup by gravity while the cup is spun by a simple electric motor. Provided the flow rate is low enough, filaments of the liquid form in the grooves without flooding over. Such equipment can produce a very uniform size spectrum of drops. Furthermore, the size of the drops can be adjusted by varying the flow rate of the liquid and the speed at which the cup is rotated. Because large drops (which use up so much liquid) are not produced, one litre of liquid will cover a hectare or more. This makes it economical to use oil rather than water as a diluting agent; using oil decreases the rate of evaporation of small drops and also increases their mass so that they behave like slightly larger aqueous ones.

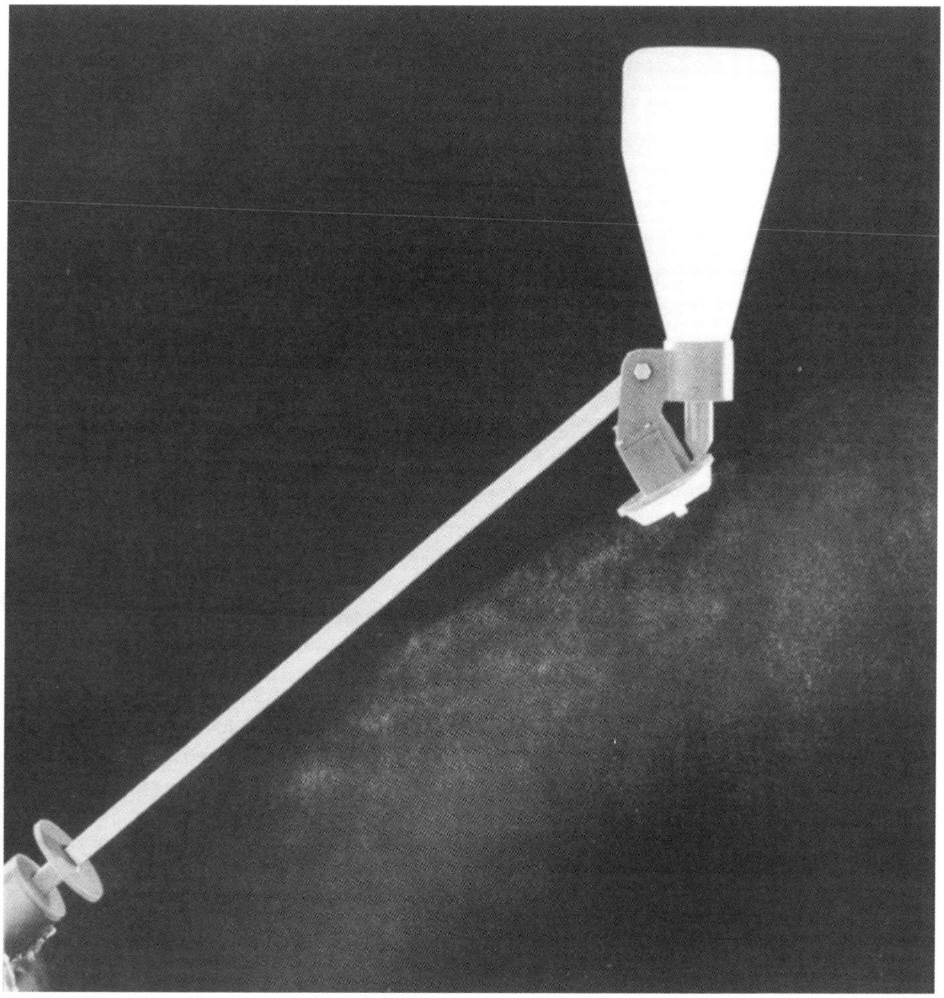

Figure 6.2 The spinning cup.

6.4.3 The electrostatic sprayer

The third approach has been to use the fact that the crop is 'earthed' to make it pull electrostatically charged drops towards itself; the charge on the drops induces an opposite charge on the target, which then attracts the drops. Coverage is superb on isolated plants, for as the charge on the top of the upper leaves becomes satisfied, drops are pulled to the underside and so on down the plant. Small drops do not drift away in the wrong direction, and volumes as small as 0.5 litres ha^{-1} become possible. The idea has even been extended to use electrostatic forces to pull filaments of spray from channels drilled in the nozzle so that drops are produced by equipment with no moving parts at all! Low-volatility oil can be used as the diluting agent. Although the electrostatic sprayer has proved an excellent device for producing drops, any cross-wind displaces them in rather unpredictable ways. Also, unless the nozzle is kept for long enough over each plant (which is rather impracticable), the drops are attracted only to the upper parts of the crop canopy, resulting in rather poor penetration of the crop.

6.4.4 Optimizing the settings of spraying equipment

With all types of spraying equipment it is possible to adjust settings which influence the size spectrum of drops that it produces. As an illustration of this, Activity 6.3 is concerned with how the flow rate of liquid and the speed of rotation of a spinning cup sprayer can be optimized to produce as uniform a size spectrum of drops as possible.

First it is necessary to consider two definitions of *mean drop diameter*. The **number mean diameter** is the drop size which divides the spray into half in terms of *numbers of drops*; in other words, half the drops are smaller than and half are larger than the drop with the number mean diameter. The **volume mean diameter** is the drop size which divides the spray into half in terms of *spray volume*; i.e. half the volume of the spray consists of drops smaller than and half consists of drops larger than the volume mean diameter. The nearer the ratio of volume mean diameter/number mean diameter is to one, the more uniform are the drops produced by the spray (in that half the spray by volume is in the smaller 50% of drops and half is in the larger 50%).

Activity 6.3 *You should spend up to 45 minutes on this activity.*

Table 6.2 shows the size distribution of drops produced (out of 400 measured) when a spinning cup was run at two settings. The drops are classified into various ranges of drop diameter, together with the percentage of spray volume accounted for by those drops.

Using the data in Table 6.2, estimate the number mean diameter, the volume mean diameter and hence the volume mean diameter/number mean diameter ratio for each setting. Which setting produces the more uniform spray?

Unless you have come across this sort of problem before, it will probably be far from obvious how you should proceed! In the following discussion it is broken up for you into a sequence of steps.

You first have to imagine all the drops arranged in order of size from the smallest (drop no. 1) to the largest (drop no. 400). The number mean diameter is the diameter of the 200th drop (strictly it is the average of the diameters of the 200th and 201st). The volume mean diameter is the diameter of the drop produced when half the volume of spray is in smaller drops and half is in larger drops (again, strictly we should allow for the tiny amount of spray in *this* particular drop).

The trouble with the data in Table 6.2 is that they are not in a form suitable to work with directly since it is not easy to identify the ranges in which the number mean diameters and the volume mean diameters fall; it is first necessary to convert them into **accumulated data**. This is best illustrated by showing in a new table the data for setting 1 when they have been so converted (Table 6.3).

The first line of data in Table 6.3 is as it was in Table 6.2a. The second line shows that 28 drops (i.e. 8 plus 20), accounting for 0.51% of the volume (i.e. 0.01% plus

0.50%), have diameters in the range 0–40 μm (in other words, in the range 20–40 μm *or smaller*). The third line in Table 6.3 comes from adding the third line in Table 6.2a to the second line in Table 6.3. The last line shows (not surprisingly) that 400 drops, accounting for 100% of the volume, have diameters in the range 0–100 μm.

(i) Convert the data for setting 2 (Table 6.2b) into accumulated data and present them as a table. Then check with our answer before proceeding.

From an examination of Table 6.3, it can be seen that the 200th drop must have a diameter in the range 60–80 μm. The drop for which it can be said that (virtually) 50% of the spray volume is in smaller drops and 50% in larger drops is also in the 60–80 μm diameter range. For setting 2 the comparable ranges are 150–200 μm and 200–250 μm respectively (Table 6.6 on p. 125). However, it is not possible to be any more precise than these (rather broad) diameter ranges when working from such tables of data.

(ii) The number mean diameters and the volume mean diameters can be estimated more precisely if we *plot* the accumulated data and then *interpolate* (i.e. read off the graphs *between* plotted points).

Plot separate graphs of the accumulated number of drops and the accumulated volume for each of setting 1 and setting 2 (i.e. four graphs in all). The *y*-axes should run from either 0 to 400 drops or 0 to 100%. The *x*-axes should run from 0 to 100 μm (for setting 1) and 0 to 350 μm (for setting 2), and the *top* of each range of drop sizes should be used to represent the accumulation of that range and lower ranges (e.g. points should be plotted at 20, 40, 60, 80 and 100 μm for setting 1). Finally, a *smooth curve* should be drawn through the plotted points for each of the four graphs. (You may find it easier to draw these curves if you also plot a point at the *origin* of each graph—it is, after all, reasonable to claim that *none* of the drops (accounting for *none* of the volume) has a diameter of 0 μm!)

(iii) The number mean diameters are then obtained by reading off the diameters that correspond to the 200th drop for setting 1 and the 200th drop for setting 2 (i.e. lines are drawn *across* at the 200 level until they meet the curves and then continued *down* to the corresponding drop diameters). The volume mean diameters for settings 1 and 2 are similarly obtained by reading off the diameters that correspond to the 50% volumes.

(iv) Finally, calculate the volume mean diameter/number mean diameter ratio for each setting and decide which setting produces the more uniform spray.

When a low flow rate and a high rotation speed are used (as in setting 1), the liquid stays in the grooves of the spinning cup to produce filaments which then shear into drops. With a reduced rotational speed and an increased flow of liquid (as in setting 2), the grooves flood over and a sheet of liquid (rather than individual filaments) breaks up at the rim.

Question 6.5 Which would be the better of the two settings for depositing drops on plants? (*Hint*: You have to bear in mind more than just the uniformity of the drops.)

The degree to which equipment such as the spinning cup overcomes the wastage problems of traditional sprayers can be measured by the reduction in pesticide per unit area which becomes possible. Clearly, if more of the pesticide is deposited on plants, we should be able to use less of it. Unfortunately, this has rarely been shown to be possible with the spinning cup; in general, farmers have not been over-impressed with the performance of the equipment.

Table 6.2 The number of drops produced (out of 400 measured) classified into various ranges of drop diameter, together with the percentage of spray volume accounted for by those drops, when a spinning cup was run at two settings: (a) low flow rate at high rotation speed and (b) high flow rate at lower rotation speed.

(a) Setting 1 (flow rate: 0.05 litres min^{-1}; rotation speed: 6 000 rev. min^{-1})

Drop diameter/μm	No. of drops in diameter range	% volume in diameter range
0–20	8	0.01
20–40	20	0.50
40–60	132	15.41
60–80	220	70.46
80–100	20	13.62

(b) Setting 2 (flow rate: 1.0 litre min^{-1}; rotation speed: 2 000 rev. min^{-1})

Drop diameter/μm	No. of drops in diameter range	% volume in diameter range
0–50	12	0.01
50–100	60	0.96
100–150	104	7.71
150–200	104	21.12
200–250	80	34.55
250–300	32	25.23
300–350	8	10.42

Table 6.3 Accumulated data for setting 1 (Table 6.2a).

Setting 1 (flow rate: 0.05 litres min^{-1}; rotation speed: 6 000 rev. min^{-1})

Drop diameter/μm	No. of drops in diameter range	% volume in diameter range
0–20	8	0.01
0–40	28	0.51
0–60	160	15.92
0–80	380	86.38
0–100	400	100.00

In fact, in discussing the size of drops, we have been rather too general so far. The ideal drop size for depositing pesticide on plants varies with, for example, temperature, wind speed and crop type. The enormous range of drop sizes produced by traditional spraying at least ensures that there are always enough drops of the 'ideal' size, whatever that may be on the day. To generate the ideal size with a spinning cup requires considerable sophistication in deciding which settings are appropriate for the prevailing conditions. Between the nozzle and the crop surface, the drop has to get through air moving both sideways and vertically; the crop canopy creates turbulence which completely changes the speed and direction of air movement there; even if it penetrates this layer, the drop then still has to impact on plant parts under conditions of very reduced air movement within the crop. *Nobody* could be expected to work out the implications of all this in the field. Although the rewards for getting the uniform drop size correct every time would be great, one is much more likely to get it wrong, with the result that very little of the spray (or possibly none at all) would be delivered to the crop.

Thus, although improvements on forcing liquids through holes are theoretically possible, these have gained little acceptance in practice. With the problems of calculating the influence of the microclimate between the nozzle and the point of impact, the error margin built into a wide spectrum of drop sizes is still required in practice. The spinning cage is probably as far as we can go at present in narrowing this spectrum.

6.5 The fate of pesticides

We have seen in Section 6.4 that maybe only 1.5% of the pesticide that is sprayed ends up on the plants it is intended to protect (or to kill). It is now necessary to consider what happens to the remaining 98.5%, which, of course, represents environmental contamination. Indeed, it is also necessary to consider the fate of the 1.5% itself since, as we shall see, not all of that is effective against the intended targets.

Although not discussed further here, it is extremely important that both any excess spray and the product container are correctly disposed of according to label instructions. Leakage from the disposal site can often result in a greater hazard to the natural environment than the whole of the actual spraying operation.

6.5.1 On and inside plants

From the moment the spray lands on a plant and dries to give a deposit of pesticide, that amount of pesticide is steadily reduced by a variety of processes. Some becomes dissolved in water transpired by the plant and so evaporates into the air. Some is dissolved in rainwater and washed onto the soil. Wind, carrying fine mineral particles, may abrade the surface deposit. Some of the pesticide is destroyed on the plant surface by oxidation catalysed by ultraviolet light (a process known as **photochemical oxidation**). Pesticide absorbed into the plant is subject to degradation by enzymes, although in some cases intermediate metabolites may be more toxic than the parent compound.

Activity 6.4 *You should spend up to 30 minutes on this activity.*

The amount of pesticide remaining on a leaf at various times after application is given in Table 6.4.

(a) Plot these data as percentage of pesticide remaining against time, fitting a *smooth* curve to the points (it may help to plot a point at 0 days, 100% of pesticide remaining). Try to extrapolate the curve beyond the given data to 50 days. What sort of pattern of decay does your graph reveal?

(b) What is the **half-life** of the pesticide on the leaf surface?

(c) Approximately how long will it take until less than 3% of the pesticide originally applied remains?

(d) Plot a second graph, this time of $\log_{10}$ (i.e. logarithm to the base 10) of the percentage of pesticide remaining on the leaf surface against time (again, extrapolating to 50 days). Does this graph give the same half-life for the pesticide on the leaf surface as you obtained in (b)?

Table 6.4 The amount of pesticide remaining on a leaf at various times after application, expressed as percentage of original amount deposited.

Time after application/days	Amount of original pesticide remaining/%
4	76
7	54
14	29
18	20
21	18
24	13
28	10
32	8

6.5.2 In the air

Pesticides reach the air through spray drift or through evaporation in transpired water from plant surfaces. Most of the pesticide stays in the air until destroyed by photochemical oxidation. The atmosphere is thus an effective 'incinerator' for pesticides. The kind of long-distance transport of pesticide by the atmosphere which may have occurred in the past with more persistent compounds, such as the organochlorines, is unlikely to occur with the modern pesticide groups.

6.5.3 In the soil

A fair amount of pesticide reaching the soil is lost to the air by evaporation and to surface water through run-off. In non-sandy soils, much of the remainder becomes strongly adsorbed onto mineral and clay particles. This adsorption slows down decomposition and, in some cases, the pesticide will be bound so strongly that it is virtually unavailable for degradation, particularly by hydrolysis (an important process for breaking down pesticide in the soil). The pesticide is thus kept in the upper layers of the soil and this reduces the contamination of drainage water and groundwater to negligible levels.

Humus contains chemical groups capable of reacting with pesticides (especially via hydrolysis), but soil microbes probably play the major role in the breakdown of pesticides in the soil into harmless products, by using them as sources of carbon and other elements. It appears that they are capable of dealing with amounts of pesticide well in excess of those applied in agriculture. Apart from in the case of organochlorines, this process seems to be completed well within a year of the pesticides arriving in the soil.

Crops and weeds also play quite an important role in taking up and thus removing pesticides from the soil. Some will eventually reach the air through transpiration, but many pesticides are chemically degraded within the plants by enzymes.

6.5.4 In water

The most serious source of pesticide contamination of water is surface run-off from treated soil, although insecticides and herbicides may be intentionally applied to lakes and ponds for pest and weed control. A little pesticide may also reach surface water from the atmosphere in rain, though this probably applied more to the organochlorines in the past than to modern pesticides.

In the water, there will be considerable dilution as well as some evaporation. Hydrolysis will occur very actively, and water-borne microbes will degrade pesticides in much the same way as do those in soil. Since water is penetrated by ultraviolet light, pesticides will also be subject to photochemical oxidation there. However, some of the pesticide will be rendered unavailable to these processes by adsorption onto suspended mineral particles, many of which may then be deposited as the bottom sediment of rivers, lakes, etc. Here conditions are anaerobic and degradation by microbes can be quite rapid, even for persistent compounds. Other pesticides accumulate in larger aquatic organisms. This is the major problem, especially with persistent pesticides such as the organochlorines, where biomagnification of residues up the food chain occurred. Only a few of the more persistent modern insecticides and herbicides can be detected in water, and even these are in minute traces (i.e. a few parts per billion). In the oceans, traces of pesticide are extremely hard to find; they probably arise from rainfall and are soon eliminated or adsorbed onto the sediment at great depths.

6.5.5 Within organisms

All organisms, e.g. insects, fungi, weeds, birds or humans, possess mechanisms for limiting the potential toxicity of foreign compounds. Their surfaces can act as barriers to the penetration of pesticides. They can lock-up pesticides in a range of storage sites out of contact with the biochemical site of action of the toxin, e.g. in animals' fat reserves. They can eliminate the toxin by excretion or secretion. They also possess enzyme systems for metabolizing foreign compounds or inactivating them by binding with them; furthermore, the activity of these systems is often increased on exposure to toxins. For some organisms (particularly fungi and weeds) there is also the possibility of using alternative biochemical pathways if one is blocked by the action of a toxin.

Killing an organism with pesticide can thus be likened to trying to fill a container that has many holes (the loss mechanisms) through a very narrow opening (the penetration barriers). The organism will be killed only if molecules of toxin arrive at the site of action faster than they are removed.

All this has two important implications concerning the effects of pesticide application. The first is that non-target organisms (including humans and wildlife) have considerable ability to eliminate pesticides safely even if they are encountered regularly, provided the doses received are small. Much is made of the *potential* hazards to humans of traces of pesticide in food and water; however, whilst the toxicity of an active ingredient may seem to be the most critical element in assessing the risk associated with a pesticide, it should be appreciated that other factors have considerable influence on the expression of this activity and hence the hazards arising from the pesticide's use. The type of formulation, the concentration of active ingredient (is it ready-for-use or must it be diluted?), the method of application, the amount to be applied, the degree of exposure experienced by the person applying the pesticide or by other species at the time of treatment and thereafter, all play a major part in determining the safety of a product in use. These principles are adopted in the *WHO Guidelines to Pesticide Classification by Hazard* (1988), which lists over 600 pesticides.

The second implication is that the target organisms are not themselves 'sitting ducks'; the potential is there for surviving pesticide treatment, even of pesticides not yet invented! Variability in the expression of this potential (whether it is, for example, resistance to penetration or an ability to increase the concentration of detoxifying enzymes) provides opportunities for natural selection to operate whenever a pesticide is used. **Resistance** to the pesticide spreads in the population, since the next generation is produced by those with the ability to survive application of the pesticide.

Activity 6.5 *You should spend up to 15 minutes on this activity.*

Figure 6.3 represents, in a general way, an increase in resistance to a pesticide in a pest population between one generation and the next. Examine the figure and read its caption carefully.

(a) *Describe* in a paragraph the important features of the figure.

(b) In another paragraph, *explain* the change between the two generations in terms of natural selection.

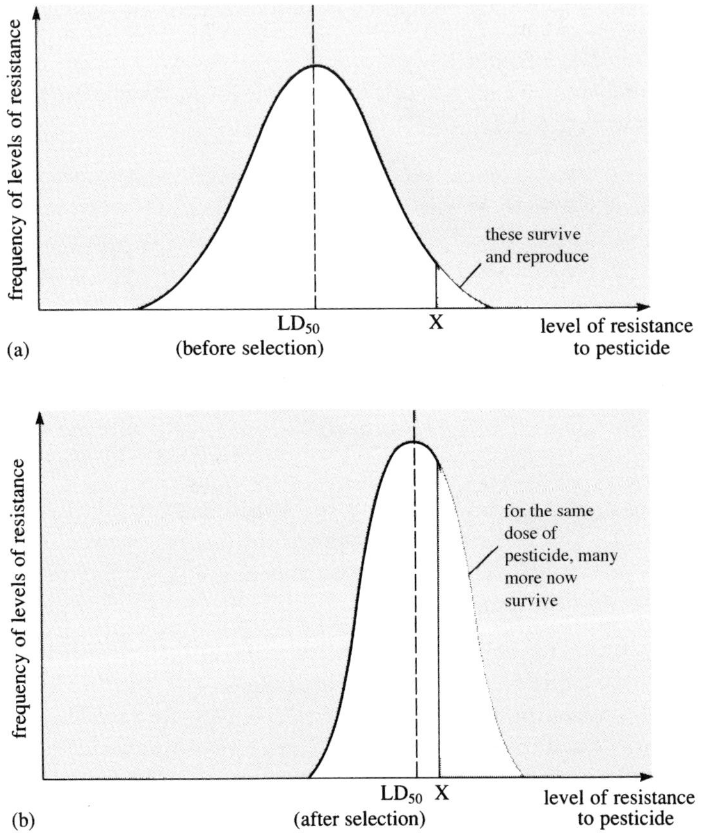

Figure 6.3 Selection for resistance to a pesticide in a pest population. (a) Distribution of resistance before selection. (b) Distribution of resistance after selection. The dose of pesticide applied to each generation is marked X; LD_{50} is the dose causing 50% mortality.

The development of resistance to pesticides can be very rapid where generation times are short (e.g. in fungi). Resistance has appeared in all types of target organism, but particularly in insects and plant pathogens. Such resistance limits the use of many of our current major pesticides, because it means that higher and higher doses have to be applied (which is costly both in an economic and an environmental sense). Moreover the appearance of resistance to pesticides is now occurring at a rate faster than the development of new ones.

Question 6.6 Why might the pesticide which gives the best kill of the pest *not* be the most desirable one?

Coping with the challenge of resistance to pesticides in the target organisms is the most important reason for modifying and reducing pesticide use. The vulnerability of pests to pesticides is a resource which should be conserved, not squandered, and all control procedures should be designed with this objective in mind. The discriminating use of pesticides whose effects are transient has many virtues and should not be dismissed, although it is often relatively costly due to the need for repeated applications. Before any spraying is carried out, consideration should be given to the resistance status of the pest population, either by direct measurement or by inference from the results following previous treatments.

Another reason for limiting the use of pesticides is that they often destroy pests' natural enemies. There are many well documented examples of this happening with insecticides. When the natural biological control of a pest is destroyed, an important restraint on how fast the pest population recovers is removed and repeated insecticide applications may become necessary. Even more serious is the destruction of the natural enemies of *other* insects, which then achieve pest status themselves and so need to be controlled. Many of today's insect pest problems are in this sense 'human-induced'. For example, a hundred years ago just one species of citrus scale-insect caused serious problems; we have succeeded in increasing the number of such problem species to nearly 70!

▷ Don't natural enemies of insect pests also become resistant to insecticides?

▶ A few natural enemies have already developed resistance to insecticides. However, an important difference between carnivores (the enemies of insect pests) and herbivores (the insect pests themselves) is that the former do not encounter plant compounds, and hence do not evolve arsenals of detoxifying enzymes, to the same extent as herbivores.

The development of pesticide resistance in the pests and the destruction of natural enemies together create a 'pesticide treadmill' whereby, having started using pesticides, there is great danger of having to use ever increasing doses and frequencies of application.

Activity 6.6

It is very important that farmers should avoid routine reliance on pesticides and integrate reduced pesticide use with other control measures. List the reasons for this in order of importance.

Can the pesticide industry ever guarantee that their products will not cause long-term damage to the environment?

No; some risk is, of course, inevitable when new compounds are introduced into agriculture. By definition, *new* compounds in the environment may present unforeseen problems, however thoroughly they are tested. This does not mean that we should stop developing new pesticides, but vigilance in their use should always be a priority.

Summary of Chapter 6

1 Agricultural monocultures are less stable than natural ecosystems and can be exploited by organisms (pests) which can significantly reduce the quality and yield of a crop.

2 If the use of pesticides were stopped, the price of agricultural commodities would rise and their quality would deteriorate.

3 The development of modern pesticides is very expensive. It involves careful evaluation of hazards, especially those which have been revealed for other pesticides in the past. Governments approve the use of new pesticides after examining the safety and other data submitted by the manufacturers and other agencies.

4 Most insecticides are nerve poisons; their diversity relates more to how they reach the insect (routes to target) than to a variety of biochemical modes of action. The diversity of fungicides and herbicides is largely based on a variety of biochemical interactions with the target organisms.

5 Modern synthetic pesticides (and their breakdown products) are much less persistent in the environment and less dangerous to humans than many of the early ones, particularly organochlorines like DDT.

6 The application of pesticides by spraying is a very wasteful process because a wide range of drop sizes is produced. Larger drops account for much of the pesticide applied, but are less likely than drops of intermediate size to be deposited on plants effectively. Very small drops are likely to be carried away from the target by air movement, etc.

7 The spinning cage, the spinning cup and the electrostatic sprayer are among the devices that have been developed in attempts to improve the efficiency of pesticide application; all present new problems which limit their usefulness in practice. The spinning cage is probably as far as we can go at present in narrowing the spectrum of drop sizes.

8 The environment is remarkably resilient to pesticides; physical and chemical processes seem able to inactivate or degrade the quantities of pesticide that currently enter it.

9 Target organisms are well adapted to withstand pesticides; it is easy for resistance to pesticides to reach levels which limit their ability to control pest populations.

10 Pesticides can destroy the natural enemies of pests (making them an even greater problem) or the natural enemies of relatively harmless species (which then become pests).

7 Integrated farming systems (2 hrs)

7.1 Introduction

'Ever-present Scandal of Farm Policies'

'Assault on Food Mountains'

'Farming takes a Heavy Toll of the Countryside'

'Vandal Farmers'

'Searching for Peace on Farm Trade Front'

'Uphill Road to Worldwide Farm Reform'

'Farmers Facing a Hazy Future'

These are a selection of headlines from articles published in the broadsheet newspapers during the late 1980s. Agriculture seemed to be in crisis and was being attacked on several fronts.

Activity 7.1

From your study of this book, the headlines above and personal experience, what do you consider to have been the main areas of concern in the UK about modern farming methods and the effects that they were having on the countryside in the past twenty years or so? Your list should include no more than 10 points.

Concerns, reflected in the headlines above, fell into three broad categories. First, there was what is now the European Union (EU) and the problem of the Common Agricultural Policy (CAP), with reports that butter 'mountains', beef 'mountains' and wine 'lakes' were costing the taxpayer millions to maintain. Then, there was the environmental lobby, unhappy about the effects that increasing use of pesticides and fertilizers were having on the safety of food and water, and on biodiversity (that is the number of plant and animal species in a given area). Finally, there were the reports from the GATT (General Agreements on Tariffs and Trade) talks. These talks never seemed to be resolved, partly because the CAP presented a barrier to the liberalisation of world trade.

Modern intensive **conventional agriculture** has become a technical and mechanized industry which benefits from economies of scale, has low demands on labour and high demands on external inputs of energy and agrochemicals, such as pesticides, fertilizers, growth promoters, etc. Alternative approaches to modern agriculture were briefly touched upon in Chapter 1. **Organic agriculture** depends entirely upon the use of natural, ecological mechanisms to maintain fertility and control pests without any external inputs. In between conventional and organic agriculture, there is the approach known variously as 'integrated', 'sustainable', 'low-input' or 'less-intensive' agriculture. It has been defined (by Jordan and Hutcheon, (1993), *Proceedings of the Fertilizer Society*, No. 346) as:

> *An holistic pattern of land use, which integrates natural resources and regulation mechanisms into farming practices to achieve a maximum, but stepwise replacement of off-farm inputs, to secure high quality food and to sustain income*

It is this **integrated agriculture** approach that we will explore in some detail in this chapter. The adjective 'integrated' is the one most often used to identify this type of farming system and also components of the system such as 'integrated pest control' or 'integrated crop management'.

Integrated farming techniques are outlined in Section 7.3 and an example of how the techniques can be put together into a farming system is described in Section 7.4. Possible future developments are briefly described in Section 7.5. Throughout, the emphasis is on arable agriculture.

First, we shall consider the political and economic factors which have contributed to the development of conventional agriculture and to the current winds of change.

7.2 Economic and political change

As you read in Chapter 1, agricultural production has increased severalfold since the Second World War. Initially, UK agricultural policy was determined mainly by the government but, since joining the EEC in 1973, it has been implemented through the CAP. We will look at the effect the CAP has had on conventional agriculture, how changes have been brought about by recent reforms, and finally how commitments to sustainable development and liberalisation of international trade, through the GATT talks, have influenced agricultural policy.

7.2.1 The Common Agricultural Policy (CAP)

The objectives of the CAP, as stated in *Article 39* of the *Treaty of Rome*, are fivefold:

- to increase agricultural productivity;
- to secure food supplies;
- to stabilize markets;
- to ensure a fair standard of living for the farming community;
- to assure reasonable retail prices for consumers.

These objectives have been achieved mainly by maintaining predetermined support prices for produce. If there were surpluses, which under free market forces would have lowered the selling price, produce was bought by the authorities at the support price—the process of intervention. Produce, which was thus effectively removed from the market, was then stored, sold to non-EU countries or distributed free to the needy. To further protect EU farmers, imports from non-EU countries were, where necessary, levied to increase their cost and ensure that they did not undercut the support prices. Under this policy, the more farmers produced, the greater their income. Production increased but the size of the market did not, so food commodities began to accumulate into large surpluses.

Technical improvements facilitated increases in production. For example, wheat yield more than trebled from about $2\,\mathrm{t\,ha^{-1}}$ in the 1940s to about $7\,\mathrm{t\,ha^{-1}}$ by the 1990s and milk yield increased from about 6 litres $\mathrm{cow^{-1}\,day^{-1}}$ to about 14 litres $\mathrm{cow^{-1}\,day^{-1}}$ over the same period. Public concern grew as the food 'mountains' absorbed 20% of the CAP budget in storage costs and 28% in export subsidies (grants). A notable example is that of butter which was sold very cheaply to the Russians. The general public considered it scandalous that the price of European butter was less in Moscow than in the UK.

Reforms to the CAP were first introduced in 1984 with milk quotas aimed at reducing the size of the powdered milk 'mountain'. The reforms were greatly extended in 1992 with a shift away from support prices to direct subsidies to farmers. The arable and beef sectors were particularly affected. For the first time support was given to measures to enhance the environment.

In order to limit surpluses of arable crops, support prices were reduced by 30% over three years starting in 1993. To compensate for this reduction, farmers received direct payments provided they set-aside 15% of their land from food production as part of their crop rotation. It was estimated, in 1993, that a farmer with 800 ha could receive up to £142 000 in compensation for the low price of crops and £25 000 for putting land into set-aside, in addition to income from crops. To some extent these reforms have taken away the incentive to overproduce as they give direct subsidies rather than payments for commodities. Farmers have also been encouraged to look to the real markets for returns on their produce. There is, however, a body of opinion which finds the distribution of payments for doing nothing with land an unsatisfactory solution.

The 'Agri-Environment Regulation' in the 1992 reforms requires all member states of the EU to introduce a programme of schemes to encourage environmentally beneficial farming methods. In the UK there has been support for a number of schemes including the Nitrate Sensitive Areas scheme discussed in Chapter 4. The Environmentally Sensitive Areas scheme, which offers farmers incentives to farm in ways designed to enhance or protect valuable landscapes and wildlife habitats, is the largest, covering 2.7 million hectares at the time of writing. The UK plans to spend a total of £100 million on the schemes and up to 75% of this can be claimed from the CAP budget. This represents about 2% of the UK's annual budget from the CAP, a percentage which many environmentalists consider to be far too low.

7.2.2 Sustainable development

The concept of **sustainable development** was launched when the World Commission for the Environment, chaired by G.H. Brundtland, published its report *Our Common Future* in 1987 and defined it as:

> *Development that meets the needs of the present without compromising the ability of future generations to meet their own needs.*

It is noticeable how even this most widely accepted definition is centred entirely on *human* activity and survival. The need for protection and maintenance of the world's natural resources for their own sake is not addressed at all.

In 1992, at the UN Conference on Environment and Development in Rio de Janeiro, the UK agreed to the framework plan for further action on sustainable development. Since then, in *Sustainable Development: the UK Strategy* (Secretary of State for the Environment, 1994), the government has stated how this is to be achieved in the UK. For agriculture, it is through the environmental schemes brought in through the 1992 reforms of the CAP, with further provision of advice, research, monitoring and, where appropriate, regulation. For the future, the report states that:

> *The government will encourage environmentally sensitive agriculture, and will work for further CAP reform to reduce levels of support and integrate fully environmental considerations.*

7.2.3 The General Agreements on Tariffs and Trade (GATT)

GATT was initiated in 1947 with the objective of reducing tariffs (that is, taxes on imports and exports), and other barriers to trade, in order to make international commerce fairer. There have been numerous rounds of talks; the latest, the Uruguay round, started in 1986 and was the first to include agricultural products. Before the

1992 reforms, the CAP was a bone of contention in view of the special protection given to EU farmers; this was one of the many reasons why the Uruguay round took seven years to complete.

The agreement, reached in December 1993, has a number of implications for the CAP. It endorsed reduced levels of support for agriculture (both tariffs and subsidies were to be cut) and EU farmers are now therefore exposed to greater competition from world agriculture. Commitment to the agreement is likely to require little adjustment to the CAP at present, but EU farmers will find themselves increasingly under pressure to meet the demands of the market place by becoming more efficient and reducing the selling prices of their produce. Viable integrated farming systems must be able to meet these demands.

It is important to note here that almost all political and economic changes in UK agriculture are now achieved through the reforms in the CAP. We have seen this in the implementation of the government's commitment to sustainable agriculture and to GATT. Financial incentives provided through the CAP are also crucial factors if farmers are to be induced to adopt integrated farming techniques and switch to integrated farming systems.

7.3 Integrated farming techniques

Modern varieties of crop plants are high yielding so long as they remain healthy and are well supplied with water and nutrients. However, they tend to give low yields if these conditions are not met. Conventional agricultural practices therefore aim to maintain healthy, fast growing plants. Fertilizers and pesticides are applied in managed programmes intended to prevent *potential* problems. For example, wheat can be treated with up to 13 different types of pesticide in six or seven applications in a single growing season.

Not surprisingly then, techniques employed in integrated farming systems concentrate on ways of maintaining healthy crops, well supplied with nutrients, whilst reducing inputs of agrochemicals. A particular theme is the desirability of targeting agrochemicals onto affected areas within a field only *after* a problem has been recognised as being economically damaging. Other equally important approaches are the implementation of good husbandry practice and the use of natural control systems to control pests and diseases.

It is important to recognise that single measures usually have multiple effects. For example, re-establishment of a hedgerow to contain stock can provide shelter for predators of pests, act as a windbreak, reduce soil erosion and enhance the appearance of the countryside.

7.3.1 Crop diversification

Cropping patterns in conventional arable farming consist of continuous winter wheat or rotations which include the alternative or break crops, winter barley and oilseed rape. If you live in an arable farming area you may have noticed few crops other than cereals and oilseed rape. Problems arise with such a mono-culture, or near mono-culture, because crop residues build up in the soil and harbour **propagules** (that is, any part of an organism that is capable of growing into a new organism) of pests, e.g. fungal spores and slug eggs. Diseases are usually specific to a particular crop and some propagules, particularly resting fungal spores, can survive for more than one season. So, if a crop is grown continuously, propagule density can build up and present a considerable source for reinfection each year. However, the survival of propagules is very much reduced during the period of growth of a non-host crop, so risk of disease can be reduced by using more diverse rotations which include at least four different crops. Crop residues can add to the organic matter of the soil and leguminous residues may improve the soil nitrogen status.

▷ What is the danger of adding leguminous residues to soils of moderate to high nutrient status?

▶ Nitrate may be leached from the soil during the following winter, especially if the plant cover is sparse (Chapter 4).

Cover or catch crops (see Chapter 4) are traditionally grown between two main crops when the soil would otherwise be bare. Their roots mop up residual nitrate very effectively (thus reducing leaching) and they provide ground cover which helps to reduce soil erosion. They are especially useful when grown during the winter, after harvest of the main crop in autumn and before a crop is sown in the following spring. After harvest, they may be sold or used as livestock food; alternatively they may be incorporated into the soil as a green manure.

Less conventional techniques for crop diversification include growing crops in alternate rows, strips or mosaics, a method successfully employed by farmers in Africa and other developing areas. Such mixtures can deter pests from host species as the presence of neighbouring crops may act as camouflage, a trap for the pest, a refuge for a predator or the odour may act as a repellent. For example, gardeners often intercrop carrots with onions to reduce damage from carrot fly, as the smell of onions is thought to deceive the fly.

An important factor that influences the level of damage from pests is the crop variety. If the breeding of varieties with high yield is the priority, natural defences to disease can be inadvertently lost. Sometimes genes for disease resistance can be bred into varieties but they may provide only short term protection against, for example, fungal disease. All too often within crop mono-cultures, strains of fungi develop that are not affected by the presence of 'disease resistance genes'; once established, such diseases can quickly spread through vast areas. One approach is to grow a variety which is made up of several populations of plants which are similar except for their type of disease resistance. Thus, while small populations of susceptible plants may become infected with disease, it is more difficult for the disease to build up within the crop as a whole.

So breeding crops for disease resistance can be a race to keep ahead of mutation in the fungi — almost a battle of wits between the microbe and the breeder. The new techniques of genetic engineering allow the introduction of genes for resistance from almost any organism and create 'designer crops' which are resistant to disease and are also high yielding, even when grown in soils of moderate to low nutrient status. Resistance is likely to be longer lasting because a foreign gene will be an unknown entity to the disease-causing organism and present a completely new challenge to be overcome. However, it is likely to be a considerable time before a range of 'tried and tested' genetically engineered varieties is commercially available.

7.3.2 Methods of cultivation

In the traditional method of cultivation — ploughing — topsoil is inverted and material, such as organic matter and associated organisms on or near the soil surface, is buried deeply. This practice is beneficial because the soil is broken up as a preliminary to seed bed preparation and harmful organisms are buried. However, there are several undesirable effects: there is a risk of erosion from the bare soil; if the soil is warm there is also a danger of nitrate leaching and volatilisation of ammonia; beneficial organisms, such as predators of pests, are buried; and, finally, an impervious layer of soil (called the 'ploughpan') may develop just below the depth of ploughing and this can impede drainage. Risks of erosion and loss of nutrients are also greater if there is a long gap between the time when the soil is ploughed and the next crop sown.

An alternative and more recent option, known as **minimal cultivation**, is not to cultivate the soil at all but to sow the seeds into drills cut in otherwise undisturbed soil. The undesirable effects of ploughing do not apply, but the presence of crop residues on the soil surface can increase the severity of damage caused by animal pests and diseases and also the amount of weed growth.

A third option, **reduced cultivation**, may therefore offer the best compromise. The soil is disturbed in order to incorporate 70% of crop residues, so the surface layer is cultivated but the topsoil is not fully inverted as in ploughing. The risks of soil erosion and loss of nitrogen are less than with ploughing and the majority of harmful organisms are buried. However, beneficial organisms including earthworms, whose food source is crop residues, and predators of pests remain near to the surface.

The effect of the method of cultivation on the incidence of one particular disease — Barley Yellow Dwarf Virus (BYDV)—is illustrated in Table 7.1. Many virus diseases of crops are transmitted by aphids which pass the virus from plant to plant when feeding on sap. Winter barley is susceptible to infection by BYDV in autumn when winged aphids migrate from grasses into seedling crops.

Table 7.1 The effect of methods of cultivation on the incidence of BYDV and the yield of winter barley.

Method of cultivation	Incidence of infection/%	Yield/t ha^{-1}
ploughing	57.3	2.2
reduced cultivation	24.3	4.2
minimal cultivation	2.7	5.1

Question 7.1 Describe, and then attempt to explain, the data in Table 7.1.

For this particular disease, therefore, minimal cultivation appears to be the most beneficial. It is thought that winged aphids are more likely to settle on ploughed fields, because the crop stands out better against the soil, than on non-ploughed fields, where the contrast is less pronounced because of the presence of surface residues. Minimal cultivation will also maintain numbers of beetle and spider predators which are thought to control the population of aphids and so reduce the spread of the virus. Minimal cultivation can, however, encourage other pests such as slugs, wireworms and leatherjackets.

7.3.3 Insect control

The question of insecticide application is discussed in some detail in Chapter 6. In this short section some of the methods which allow reductions in the levels of application will be considered. There are two approaches, the first targets insecticide applications so reducing the need for blanket sprays over the whole crop and the second encourages the natural predation of pests.

Targeted application

The use of insecticides can be reduced by more precise targeting of chemicals. Preventative blanket applications can be avoided by applying knowledge of economic thresholds for cost effective control, knowledge of the ecology of pests and regular monitoring of the crop for signs of damage. In the case of BYDV (Section 7.3.2), the disease is routinely controlled by applications of broad spectrum insecticides which aim to kill the aphids, but are likely to kill a whole range of other insects as well. Incidence of the disease is, however, spasmodic and computer simulation models are available which can be used to forecast quantitatively and reliably the risk of incidence of BYDV. Insecticides need then be applied only when the risk level warrants it.

Natural predation

Biological control, where an exotic predator is introduced to control a pest, such as an insect, has rarely been found to be successful in the field. These techniques have, however, been applied more successfully in glasshouses where, for example, populations of whitefly (*Trialeurodes vapourariorum*) can be controlled by the parasitic wasp *Encarsia formosa*. In field crops, attention is increasingly being focused on the ability of natural populations of predators to control insect pests, especially aphids. There are two main groups of predators: the *polyphagous* insects, such as beetles, which consume many types of prey; and the *aphidophagous* insects, such as ladybirds, hoverflies and parasitoid wasps, which prey only on aphids.

In the 1980s, a group of researchers at Southampton University used infrared cameras to study the predation of aphids in wheat fields. Some beetles, like the ground beetle *Bembidion*, were observed to make straight for patches of aphids and concentrate their feeding on them. From these observations, further work concentrated on techniques to aid the survival of the beetles and bring them within short 'walking' distances of their prey. The result was 'beetle banks'—raised banks at least one metre wide which subdivide large cereal fields. They harbour beetles and, from their central position in the field, predators can reach their prey within hours rather than days. Banks are constructed by mounding up soil and sowing grass mixtures which include tussock grasses such as cocksfoot (*Dactylis glomeratus*) and Yorkshire fog (*Holcus lanatus*). They provide shelter for the predators during winter as the variation in temperature within the tussocks is much less than in the crop (although the average annual temperatures are similar). Studies showed that populations on the banks exceeded 100 beetles m^{-2}, five to ten times more than in natural boundaries. Where the technique has been put into practice, farmers find that the cost of bank establishment is negligible compared to the benefit. However, other farmers just regard the banks as obstacles to cultivation.

Being winged, aphidophagous insects can travel moderate distances to reach their prey. They lay their eggs in vegetation at times when prey is scarce and, as they are on site when the prey first appear, they are thought to be able to prevent populations of aphids from building up. In spring, they have a special need to consume sufficient food rich in types of protein which will ensure that their eggs will mature. These proteins are found in pollen and nectar from spring flowers, particularly members of the family Umbelliferae, such as cow parsley (*Anthriscus sylvestris*), that have open flowers with easy access to insects. Methods of integrated crop management aim to provide such flowers when and where they are required.

▷ Give three possible ways of increasing wild flowers in and around arable fields.

▶ Plant wild flowers at the edges of fields.

Allow wild plants to grow within the crop.

Establish conservation headlands, i.e. edges left unsprayed with pesticides. Reduced herbicide use will enhance the survival of wild flowers and reduced insecticides will enhance the survival of pollinating insects.

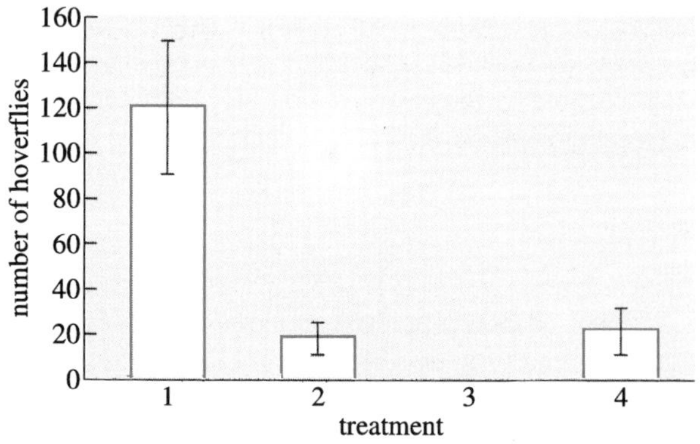

Figure 7.1 Numbers of adult hoverflies counted in field margin plots with the following treatments: (1) a wild flower and grass mixture; (2) regenerated natural vegetation; (3) bare ground; (4) a grass mixture. The field margin plots were established in September 1991 and numbers of hoverflies were counted in August 1992. The bars represent the standard error of the mean, a measure of variability that has similarities with the standard deviation.

If wild flowers are planted they can be native or introduced species. Currently, a particular favourite is *Phacelia tanacetifolia*, a North American species, which has cheap seed, establishes well and competes favourably with weeds. It has a long flowering season and is attractive to hoverflies and bees. It is also susceptible to herbicides and can be checked if it gets out of control. Because *Phacelia* is so easy to grow it can also be used as a cover crop (Section 7.3.1). Evidence that the presence of wild flowers can increase the numbers of aphidophagous hoverflies is presented in Figure 7.1. The establishment of grasses and the regeneration of natural vegetation had little effect on the population density of hoverflies; numbers built up only where wild flowers were grown.

The question then is, how effective are the predators once their population densities have increased? In a second experiment, three fields bordered with *Phacelia* were compared with three control fields without flower borders. When the numbers of wheat stems with one or more aphids were counted, the results were not so clear cut (Figure 7.2). The infestation of wheat by aphids was not affected by the presence of the *Phacelia* borders until early July, four weeks after the first appearance of aphids in the crop. However, during July there was a measurable level of control.

▷ Using the data in Figure 7.2, what was the approximate range of infestation on 12.7.93 in (a) the control fields, and (b) the fields bordered with *Phacelia*?

▶ (a) 45% to 50% of wheat stems were infested with aphids in the control fields, (b) 25% to 40% of wheat stems were infested in the *Phacelia* fields.

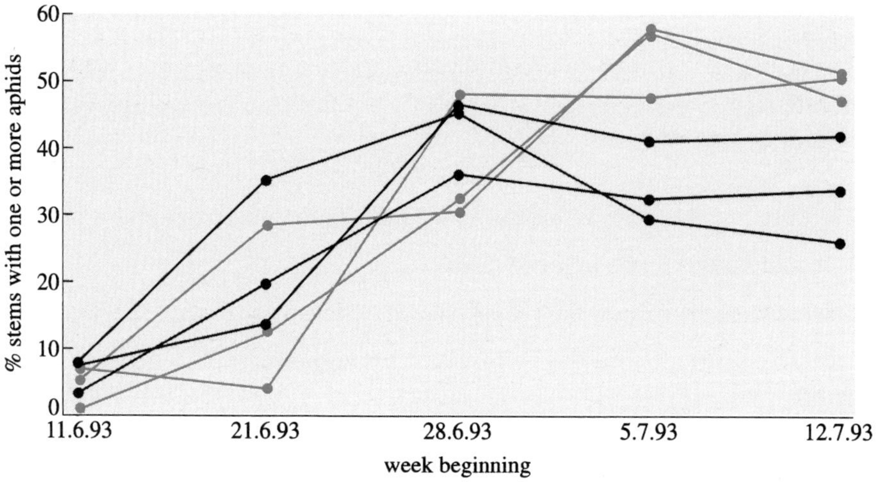

Figure 7.2 The percentage of wheat stems infested with aphids from three control fields (marked with coloured circles) and from three fields bordered with *Phacelia* (marked with black circles) over four weeks in the summer of 1993.

Further investigation showed that the control of aphids coincided with the appearance of the third and hungriest growth stage of hoverfly larvae. So, to be effective, not only do aphidophagous insects have to be present, they also need to be at an appropriate growth stage. Research continues to seek ways to manipulate natural predation for pest control under agricultural conditions.

7.3.4 Disease control

A major cost in wheat production is that of fungicides. Surveys during 1990/1991 showed that 24% of wheat crops were given between 3 and 7 sprays of broad spectrum fungicides during the season as part of managed disease control programmes.

Many diseases build up in the soil and in crop residues and are influenced by the number and sequence of susceptible crops in the rotation. Diseases can be avoided by the selection of appropriate resistant varieties, break crops, cultivation methods and sowing dates (Section 7.3.1). Crops which have received high applications of nitrogen fertilizer are also more prone to disease through changes in the structure of the crop canopy and increased susceptibility of leaf tissues to infection.

Regular examination of the crop for fungal disease can reduce the need for preventative applications of fungicide. This information, combined with an understanding of the levels of infection and the conditions under which diseases can adversely affect crop yield, can cut rates of application by up to 85% while still maintaining yields. The main drawback with this approach is that farmers often lack the knowledge and skill to be able to recognise the early stages of disease symptoms.

7.3.5 Weed control

The successful implementation of integrated crop management systems is thought to be most threatened by the problem of weed control. In conventional farming, weed free crops can be obtained with the use of herbicides. In less intensive arable systems, weeds can be controlled by mechanical cultivation or harrowing but these methods are successful only if carried out under suitable weather conditions and at appropriate growth stages. It is particularly difficult to destroy grass weeds in autumn sown crops under minimal cultivation.

Where chemical methods are integrated with mechanical methods, one approach is to apply herbicides that target particular weeds, require low volumes of active ingredients and hence present a minimal environmental hazard.

Weeds tend to occur in patches rather than being distributed uniformly over a whole field. Methods of patch spraying, which make use of tractor-mounted computers are currently being developed to reduce the levels of application of herbicides. The first step is to prepare a computerized field map from aerial photographs and field walking data. During regular field inspections, locations of patches of weeds are entered into the computer which can then use the information to navigate the tractor to the exact locations of weed patches before spraying. This technique offers considerable financial and environmental benefits as demonstrated by the data presented in Figure 7.3 overleaf.

7.3.6 Provision of nutrients

For growth, plants require just three inputs: energy, which they obtain from photosynthesis, and water and mineral nutrients, both of which they absorb from the soil. Six mineral nutrient ions are required in large amounts—NO_3^-, PO_4^{3-}, K^+, Ca^{2+}, Mg^{2+} and SO_4^{2-}—and fertile agricultural soils in the UK are rarely deficient in any of the minerals except for nitrate. This is partly because applications of non-nitrogenous fertilizers over the past 40 years have built up the nutrient supply and partly because the

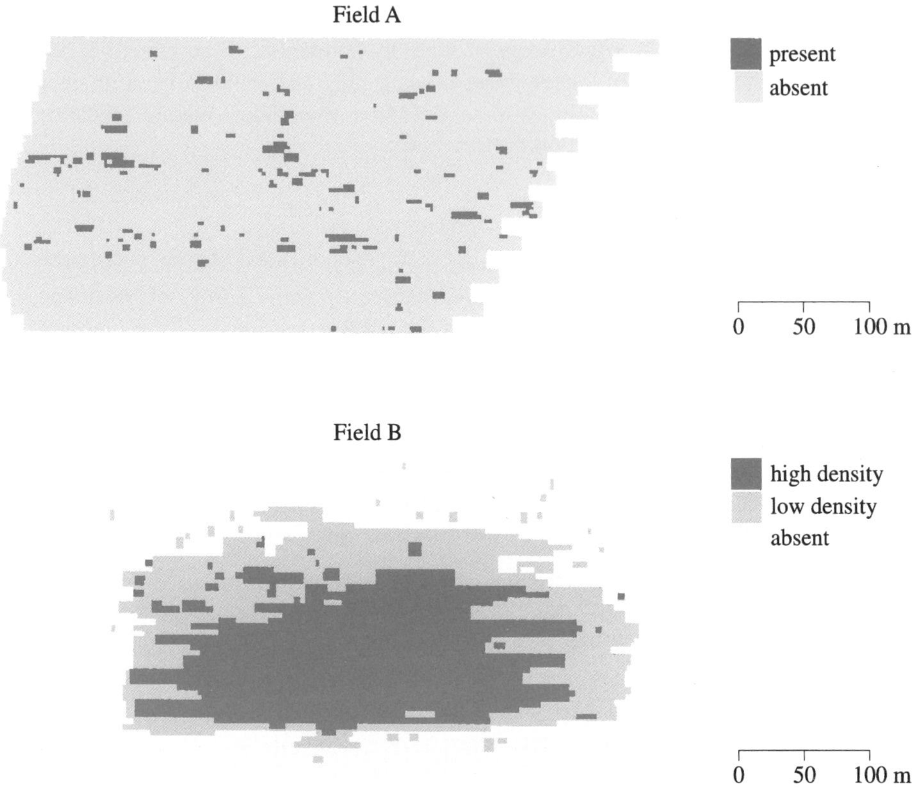

Figure 7.3 Computerized field maps showing the density of couch grass (*Elymus repens*) in two cereal fields, recorded by two people sitting on each side of a driver in a tractor travelling at 5 km hr^{-1}. The maps show the patchiness and level of weed infestation in the two fields. The repeatability of recording was found to be 75% when tested over three runs. The potential savings in herbicide application with patch spraying compared with whole field spraying were calculated to be 93% in field A with 7% couch grass infestation and 57% in field B with 51% weed infestation.

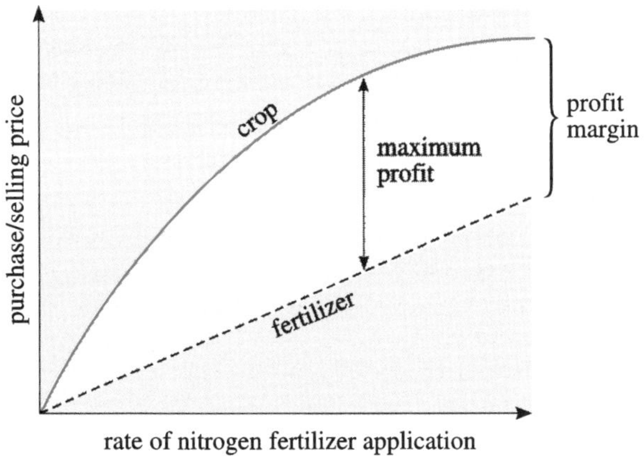

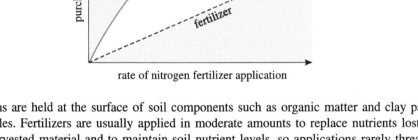

Figure 7.4 The profit margin from the application of nitrogen fertilizer. The purchase price of nitrogen fertilizer increases linearly with the rate of application, but the yield of the crop, and hence its selling price, increases curvilinearly. Hence, the profit margin, which is the difference between the purchase price of the fertilizer and the selling price of the crop, varies over the rates of application. Farmers try to apply nitrogen fertilizer to give maximum profit.

ions are held at the surface of soil components such as organic matter and clay particles. Fertilizers are usually applied in moderate amounts to replace nutrients lost in harvested material and to maintain soil nutrient levels, so applications rarely threaten the environment. Sulphate is also supplied in acid rain.

The use of nitrogen fertilizers and the 'nitrate problem' was discussed in depth in Chapter 4. Here we consider how levels of application of nitrogen fertilizer may be reduced. The conventional method of determining application levels is to measure the response of the crop to nitrogen and calculate the level at which the return on extra yield no longer pays for the cost of extra fertilizer (Figure 7.4). In Chapter 4 it was pointed out that, for winter cereal crops, the economic optimum rate of application is usually similar to the environmental optimum. Fertilizer recommendations given by the Agricultural Development and Advisory Service (ADAS) also take into account the soil type, reserves left by the previous crop and atmospheric deposition.

▷ Winter wheat is to be grown in a field following oilseed rape and requires 280 kg N ha^{-1} to achieve a yield of 7 t ha^{-1}. If the residual mineral nitrogen in the soil is about 10 kg N ha^{-1} and atmospheric deposition is estimated to be 35 kg N ha^{-1} over the growing season, how much nitrogen fertilizer would be needed to attain the target yield?

▶ Nitrogen fertilizer required = [280 − (10 + 35)] kg N ha^{-1} = 235 kg N ha^{-1}.

If excess nitrogen fertilizer is present in the soil there is a danger of nitrate leaching. This can be avoided if the total application is made in several small dressings in spring when the plant is actively growing and can rapidly absorb the fertilizer.

A more localized approach aims to apply sufficient nitrogen to meet the needs of an individual crop. Field soil is tested for residual mineral nitrogen (N_{min}, see Chapter 4), usually in spring, and recommendations are made based on the nitrogen available in a particular field at a particular time. As there is variability in the fertilizer requirements of a crop even within fields, current research aims to develop methods to target fertilizers even more specifically.

An approach which may also contribute to conserving nitrogen in soils is to apply fertilizer in the form of ammonium (which is held on soil particles), together with a nitrification inhibitor. This slows down the formation of nitrate and reduces the risk of loss by leaching.

Question 7.2 In what ways could it be said that the integrated techniques described in Section 7.3 substitute labour and knowledge for external inputs?

7.4 An integrated farming systems approach

This section describes how the techniques outlined in Section 7.3 have been combined with considerable success in the first large-scale research experiment into integrated farming systems — the LIFE (Less-Intensive Farming and the Environment) project being carried out by the Institute of Arable Crops Research near Bristol. The aim is to understand and optimize the ecological interactions within farming systems by focusing research effort on the need to resolve the conflicting requirements for control of insect pests, diseases and weeds and the need to minimize the environmental impact of crop production on biodiversity. For any integrated low-input system to be taken up by farmers it must be seen to be profitable and also to provide a reliable income from year to year. This depends to a large extent on the stability of crop yields from year to year.

In 1989, a long-term farm-scale experiment was set up on 23 hectares of land which was previously impoverished grassland. The experimental area comprised five large fields, each divided into four units of about one hectare. One of the following crop management systems was applied to each unit within each field:

System 1 conventional rotation with standard farm practice

System 2 conventional rotation with lower input options

System 3 integrated rotation with standard farm practice

System 4 integrated rotation with lower input options

Details of the rotations (conventional or integrated) and land management decisions (standard farm practice or lower input options) are given in Tables 7.2 and 7.3 overleaf. The conventional rotation of intensive cereals was chosen because, in a survey of farmers, it was found to be the most often used. Continuous cereals encourage the build up of pests and diseases and the rotation was thought to demand high inputs of pesticides and energy. The integrated rotation, which included more break crops, was chosen for a lower risk of pests and diseases. In 1993, following the reforms of the CAP, set-aside was incorporated in the rotation. It was added between the first and second wheat crops in the conventional rotation, changing the rotation from four to five years and it replaced beans in the integrated rotation. The land management decisions taken under standard farm practice were those expected to be made by a technically competent farm manager, and those taken under the lower input options were based upon the integrated techniques outlined in Section 7.3. There were five units allocated to each system—one in each of the five fields. It was therefore possible to grow each of the crops in each phase of the five-year rotation, on one of the units in each field each year.

Table 7.2 Crop rotations employed in the LIFE project.

Conventional rotation		Integrated rotation	
pre-1993	1993/1994	pre-1993	1993/1994
winter wheat	winter wheat	winter wheat	winter wheat
winter wheat	set-aside	winter oilseed rape	spring oilseed rape
winter barley	winter wheat	winter wheat	winter wheat
winter oilseed rape	winter barley	winter oats	winter oats
	winter oilseed rape	winter beans	set-aside

Table 7.3 Land management decisions employed in the LIFE project.

	Standard farm practice	Lower input options
crop variety	high yielding	disease resistant
cultivation	plough	reduced cultivation
sowing date	September	October
nutrients	optimum supply	apply if indicated by soil and plant analysis
crop protection	managed control programme	apply if pests are forecast and if disease threshold is reached; control weeds mechanically and with low doses of herbicides.

To illustrate the main findings of the experiment, only data from System 1—which represents a modern conventional agricultural system—and System 4—which represents the best known practice in integrated agriculture at the time the project started—are discussed here. From now on, System 1 will be referred to as the *conventional system* and System 4 as the *integrated system.*

The average annual application rates of nitrogen fertilizer and pesticides from 1990 to 1992 are presented in Table 7.4. Differences in the amounts applied to the two systems resulted from the different land management decisions (Table 7.3). For what is considered to be the optimum supply of nitrogen fertilizer in the conventional system, an average of 166 kg N ha^{-1} yr^{-1} was applied. However, when applications were made only if indicated by soil and plant analysis in the integrated system, an average 80 kg N ha^{-1} yr^{-1} was applied; that is, just 48% of what was considered to be the optimum supply (Table 7.3).

▷ In the conventional system, applications of pesticides were made according to a managed control programme. But how were weeds, diseases and pests controlled in the integrated system?

Table 7.4 Mean annual applications of nitrogen fertilizer and pesticides in the LIFE project from 1990 to 1992.

System	Nitrogen fertilizer		Herbicides		Fungicides		Insecticides	
	/kg N $ha^{-1}yr^{-1}$	/%†	/kgAI* $ha^{-1}yr^{-1}$	/%	/kgAI $ha^{-1}yr^{-1}$	/%	/kg AI $ha^{-1}yr^{-1}$	/%
conventional	166	100	3.48	100	1.29	100	0.03	100
integrated	80	48	2.17		0.28		0.004	

† percentage of conventional

* Active Ingredient

▶ Weeds were controlled mechanically and with low doses of herbicides. Fungicides were applied if disease thresholds were reached. Insecticides were applied if serious insect pest problems were forecast (Table 7.3). An important point here is that decision making must be flexible in order to meet the needs of the particular crops.

▷ What were the application rates of herbicides, fungicides and insecticides under the integrated system as a percentage of those applied under the conventional system? Fill in the blanks in Table 7.4.

▶ Under the integrated system, application rates of herbicides were 62%, fungicides 22% and insecticides 13% of those applied in the conventional system.

Now we will look at how these large reductions in the rates of application of agrochemicals affected the yields of the crops and also the profitability, as estimated by the gross margin. The gross margin is simply the difference between the estimated revenue for the crop, taking the average market price for the UK, and the cost of inputs such as fertilizers, pesticides and fuel. For profit on a whole farm basis other factors, such as subsidies and land rental, must also be taken into account.

The mean crop yields, costs of inputs and gross margins from 1990 to 1994, that is, over each year of the five year rotation, are summarized in Table 7.5 overleaf. Standard deviations of the values that were used to calculate the means are also included. As the standard deviation is a measure of the spread of values which make up the mean, then the lower the standard deviation, the more stable the yield, costs or margins over time.

Activity 7.2 *You should spend up to 30 minutes on this activity.*

Examine Table 7.5 and answer the following questions.

(a) How do the average yields for wheat in both systems compare with the UK average yield in the 1990s quoted in Section 7.2?

(b) In general terms, how do (i) the yields and (ii) the standard deviations of the yields compare for crops under the conventional system with crops under the integrated system?

(c) Using data in Table 7.5, calculate and enter into the partially completed Table 7.6, the yields, costs of inputs and gross margins for crops in the integrated system as a percentage of these values for crops in the conventional system.

(d) In general terms, how can the percentage yields and costs of inputs be used to explain the percentage gross margins in Table 7.5?

(e) How could the crop rotation in the integrated system be adapted to reduce further the reliance on nitrogen fertilizer?

Table 7.5 Five year summary of the mean crop yields, costs of inputs and gross margins of the LIFE project (1990–1994) and the standard deviation (s.d.) associated with the means.

	Crop yield /t ha^{-1} yr^{-1}		Cost of inputs /£ ha^{-1} yr^{-1}		Gross margin /£ ha^{-1} yr^{-1}	
	mean	s.d.	mean	s.d.	mean	s.d.
Conventional system						
winter wheat 1	8.2	1.0	253	7	819	18
set-aside	–	–	–	–	–	–
winter wheat 2	8.2	1.4	318	7	636	21
winter barley	7.4	0.7	244	3	554	7
winter oilseed rape	2.3	0.8	282	11	410	26
Integrated system						
winter wheat 1	6.9	0.4	144	5	769	8
winter/spring oilseed rape	1.7	0.7	175	10	419	15
winter wheat 2	7.1	0.6	202	7	745	4
winter oats	6.0	0.7	110	5	626	14
set-aside	–	–	–	–	–	–

Table 7.6 Yields, costs of inputs and gross margins of the integrated system as a percentage of those from the conventional system in the LIFE project.

Crop	Yield/%	Cost of inputs/%	Gross margin/%
winter wheat 1 winter wheat 2 winter oats/barley	84	57	94
winter/spring oilseed rape	74	62	102

In conclusion, the integrated system resulted in substantial reductions in the rates of application of nitrogen fertilizer and pesticides when compared to the conventional system. For pesticides, there were fewer applications rather than reductions in dose. Gross margins for the integrated system were comparable to, or greater than, those for the conventional system for all but one crop. The costs of inputs were lower, as less agrochemicals were applied, and this compensated for loss of revenue due to lower crop yields. Where the gross margin was less, a disease resistant, but low yielding variety, had been grown. This result highlights the need to breed varieties for multiple characteristics such as high yield and disease resistance. Crop yields, and hence gross margins, in the integrated system were as stable from year to year or more so, than in the conventional system.

Now that one five-year rotation has been completed, the implications of the results will be carefully considered and the experimental treatments and design of the project will be changed to answer new questions that have arisen.

7.5 Future developments

The development of suitable integrated farming techniques has, over the past few years, become part of the mainstream research effort. Many projects, at all levels of organisation, are under way in addition to that discussed in Section 7.4. The key question is, can integrated farming systems really produce sufficient food and maintain farmers' incomes on a nationwide scale and still remain sustainable? The farming community is unlikely to have confidence in these systems until this question is answered to its satisfaction.

A substantial move towards integrated systems would also require a change in UK government policy and reform of the CAP. Although the 1992 reforms weakened the link between the payment of subsidies and high yields, there is little financial support for the adoption of integrated farming systems (Section 7.2). Once the technical, political and economic problems are resolved, reliable information must then be communicated to farmers. MAFF currently produces Codes of Good Agricultural Practice for the protection of soil, water and air. The National Farmers Union (NFU) has developed a scheme in collaboration with large food retailers for the promotion of methods to produce good quality, safe food at affordable prices. Husbandry protocols have been developed for individual crops which could be applied throughout the whole food industry. This is clearly an important initiative as much of today's food is sold through the big supermarkets.

An independent organisation, Linking the Environment and Farming (LEAF), which is supported by MAFF, NFU, the food industry and conservation groups, also encourages farmers to adopt integrated techniques. In addition to providing information, LEAF offers help in carrying out on-farm environmental audits. Demonstration farms run by volunteer farmers throughout the UK also welcome interested groups and show how the techniques can be put into practice.

These initiatives, involving many influential bodies promoting the latest techniques, may in time help transform the agricultural industry and possibly the whole of the countryside.

Summary of Chapter 7

1 Since the Second World War, agricultural policies of the UK government, and later the European Union, have encouraged increases in agricultural production and ensured that farmers receive a reasonable income.

2 In the 1980s, increasing concern over agricultural surpluses led to reforms of the CAP which weakened the link between production and payments of subsidies. On the completion of the Uruguay round of the GATT talks, the EU was bound, through the CAP, to comply with measures to liberalize world trade in agricultural products.

3 The reforms in the CAP, GATT and commitments to sustainable development, have all encouraged the movement to less costly, more efficient and more environmentally friendly farming systems.

4 Integrated farming systems occupy the middle ground between high-input conventional agriculture and minimal input organic farming.

5 Integrated techniques aim to reduce applications of agrochemicals and incorporate natural control systems, good husbandry methods and knowledge of ecology. There is a substitution of labour and knowledge for external inputs. Currently, a substantial research effort is being directed at developing techniques which are less harmful to the environment.

6 When integrated farming systems are compared with conventional high-input systems, yields of crops are usually lower but the costs of inputs are also lower. As a result, the overall gross margins are similar or a little higher. From year to year, the stability of yields is generally greater in integrated systems.

7 Information on agricultural practices which minimize harm to the environment is being passed on to farmers by MAFF and independent bodies such as LEAF. Large retailers have collaborated with the NFU to develop protocols for crop production based on integrated techniques.

NEXT * Read AV notes & listen to "Introducing probability" on Tape 2, side 1. (1 hour)

Further reading

Chapter 1

Simmonds, N. W. (1979) *Principles of Crop Improvement*, Longman.

Chapter 2

Scarisbrook, D. H. and Daniels, R. W. (1986) *Oilseed Rape*, Collins.

Chapter 3

Halley, R. J. and Soffe, R. J. (eds) (1988) *Primrose McConnell's Agricultural Notebook*, 18th edn, Butterworth.

Spedding, C. R. W. (ed.) (1992) *Fream's Agriculture*, 17th edn, Blackwell.

Chapter 4

Addiscott, T. M., Whitmore, A. P. and Powlson, D. S. (1991) *Farming, Fertilizers and the Nitrate Problem*, CAB International, Wallingford, Oxon.

Chapter 5

The BSE story will undoubtedly continue to unfold during the lifetime of *Science Matters*. The best way to keep up with it is to read articles in the 'serious' press and in magazines such as *New Scientist*.

Chapter 6

Carlisle, W. R. (1988) *Control of Crop Diseases*, New Studies in Biology, Edward Arnold.

van Emden, H. F. (1989) *Pest Control,* New Studies in Biology, Edward Arnold.

Chapter 7

Pretty, J.N. and Howes, R. (1993) *Sustainable Agriculture in Britain: Recent Changes and New Policy Changes*, International Institute for Environmental Development, Research Series No. 2, London.

Skills

In this section we list skills that have been explicitly taught and/or revised in this book. You should find that most of them are special instances of the general skill categories given in the *Course Study Guide*. After each one, there is a list of questions and activities where that skill is practised.

1 Describe or interpret data, including experimental results and levels of statistical significance, presented as tables, graphs, diagrams and maps. (*Questions 2.4, 3.1, 3.2, 3.6, 3.8, 3.9, 4.1, 6.5 and 7.1; Activities 1.1, 2.1, 3.1, 4.3, 5.1, 6.1, 6.4, 6.5 and 7.2*)

2 Plot data graphically and extract information from the graphs, including by extrapolation. (*Activities 3.1, 5.1, 6.3 and 6.4*)

3 Convert data between different mathematical and tabular forms. (*Question 6.4; Activities 5.1, 6.2–6.4 and 7.2*)

4 Carry out, design or critically evaluate an experiment or other scientific investigation. (*Questions 3.7 and 7.1; Activities 3.1, 4.3, 4.6 and 5.1*)

5 Summarize, in writing, the main points, including definitions of scientific terms, from a section of teaching text that you have studied. (*Question 7.2; Activities 2.2, 4.4 and 4.5*)

6 Explain biological phenomena in terms of natural selection. (*Questions 3.3 and 6.6; Activity 6.5*)

7 Distinguish between correlations and causal relationships. (*Activity 4.1*)

8 Consider social, political and ethical aspects of a scientific issue. (*Questions 3.4, 5.1, 6.1 and 6.2; Activities 4.7, 4.8, 5.2, 6.6 and* 7.1)

9 Formulate a personal opinion or strategy on a scientific issue, including an assessment of risk. (*Question 5.1; Activities 4.2, 4.3, 4.8, 5.2 and 7.1*)

Answers to questions

Question 2.1

In each case, the diploid chromosome number of the derived species is the sum of the diploid chromosome numbers of the 'parent' species.

Cabbage/kale (to take an example) has 9 homologous pairs of chromosomes (i.e. 18 chromosomes altogether). It is the orderly association of homologous chromosomes during meiosis that leads to the production of gametes, each carrying one representative of each pair of chromosomes (i.e. 9 chromosomes). At fertilization, two haploid (n = 9) gametes fuse to produce a diploid ($2n$ = 18) zygote; once again these 18 chromosomes comprise 9 homologous pairs which can associate properly during meiosis to produce viable gametes.

Now consider what would happen if gametes of cabbage/kale (n = 9) were to fuse with gametes of turnip-rape (n = 10). The resulting hybrid (having 19 *non-homologous* chromosomes) would be infertile because none of the chromosomes would have a homologue with which to pair during meiosis. However, if the chromosome number of the hybrid were to double for some reason (which occasionally happens in nature), then it would become fully fertile with a diploid chromosome number ($2n$) of 38 (i.e. 19 *homologous pairs* of chromosomes). This is believed to be how rape arose as a hybrid between cabbage/kale and turnip rape (and, indeed, how mustard rape arose as a hybrid between black mustard and turnip-rape and Ethiopian rape as a hybrid between black mustard and cabbage/kale). Hybrids formed in this way have the entire somatic chromosome complements of two species and are known as *amphidiploids*.

Question 2.2

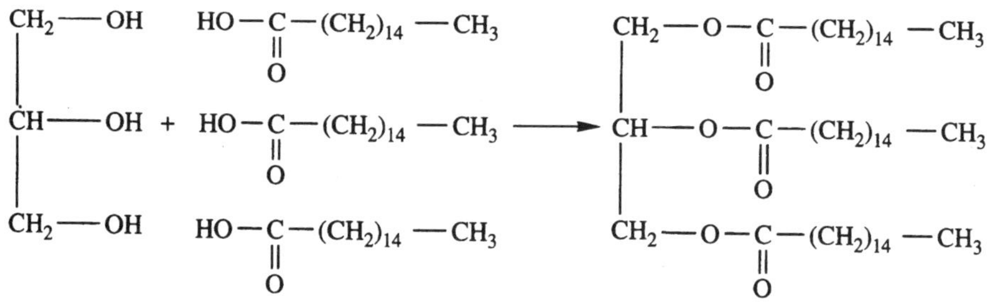

glycerol — palmitic acid — glyceryl tripalmitate

Because three hydroxyl groups (in the glycerol molecule) condense with three carboxyl groups (one in each of the palmitic acid molecules), three molecules of water are formed per molecule of glyceryl tripalmitate produced.

Question 2.3

Palmitic acid contains a total of 16 carbon atoms and no carbon–carbon double bonds; it is therefore written C16:0. Oleic acid contains a total of 18 carbon atoms and one carbon–carbon double bond; it is therefore written C18:1. Linoleic acid contains a total of 18 carbon atoms and two carbon–carbon double bonds; it is therefore written C18:2.

Question 2.4

Blockage of the step (6) between eicosenic acid and erucic acid would be the most selective means of preventing (or at least reducing) the synthesis of erucic acid. The step (5) between oleic acid and eicosenic acid would be the next most selective. Which of steps 5 and 6 you would favour depends on the usefulness of eicosenic acid.

If either of the steps leading from oleic acid to linoleic acid and linolenic acid (steps 3 and 4) were blocked, not only would the synthesis of erucic acid *not* be prevented, one or both of the unsaturated linoleic and linolenic acids would not be synthesized. Blockage of any of the steps other than 3 or 4 would reduce the amount of erucic acid. However, blockage of either of the steps leading from palmitic acid to stearic acid and oleic acid (steps 1 and 2) would prevent the synthesis of *any* unsaturated fatty acids.

Question 3.1

At hatching (or soon after), the birds experience 22 hours of artificial 'daylight' in each 24 hours. This is reduced in stages to 8 hours at 2 weeks, thus simulating the short days of winter at temperate latitudes. These 'winter' conditions are maintained for about 17 weeks (during which time the birds continue to grow), before the apparent day length is increased steadily over a 12-week period to the final laying photoperiod of 17 hours which is then maintained. This relatively rapid increase in 'day length' simulates the onset of spring, which in turn presumably stimulates the birds to commence egg laying—once a 'critical' minimum photoperiod has been reached.

Question 3.2

Lysine, threonine and sodium chloride are listed both as ingredients (in Table 3.2) and as nutrients (in Table 3.3). Lysine comprises 0.63% of the diet by weight as an ingredient and 1.4% as a nutrient. The comparable figures for threonine are 0.14% and 0.85%, and for sodium chloride they are 0.18% and 0.6%. In each case the figure for the nutrient is higher than that for the ingredient.

The most probable explanation for these differences is that there is generally more than one source of a nutrient in the diet—considering the two amino acids, as well as being supplied pure these will be present in the nutrient protein which occurs in several of the ingredients of the diet (e.g. barley, wheat, fish meal).

Did you notice that the figures in the right-hand column of Table 3.2 sum to 100% while those in the right-hand column of Table 3.3 sum to only about 40%? This implies that about 60% by weight of the ingredients do not contribute to nutrients in the diet. Most of the 'missing' 60% will be carbohydrates (i.e. energy-yielding ingredients) and some of it will be minor nutrients not included in Table 3.3.

Question 3.3

Antibiotics are employed against micro-organisms (or microbes, for short). Microbes reproduce extremely rapidly and large populations can build up very quickly. In these circumstances, there is a possible risk that a population of microbes will develop resistance to an antibiotic as a result of natural selection (i.e. through resistance arising spontaneously in maybe only one organism, whose genotype rapidly becomes predominant in the population). This immediately renders the antibiotic less effective, and therefore less useful, as a therapeutic agent. The (largely unavoidable) process of development of resistance is at least slowed by limiting how extensively an antibiotic is used. It is clearly advantageous to ourselves if there is no possibility of microbes developing resistance to the antibiotics used in human medicine through the same antibiotics being widely used for farm animals. The problem with broad-spectrum antibiotics is the danger of resistance to them (and maybe other antibiotics) developing in *several* species or strains of microbe simultaneously.

Question 3.4

You might have suggested *nutritional value* (e.g. which nutrients are present, whether the meat is fatty), *organoleptic quality* (i.e. the smell and taste) as well as the *visual appeal* of the meat.

Question 3.5

As discussed in Chapter 2, fats and oils contain fatty acid residues, which are composed of chains of carbon atoms with varying numbers of double bonds. Fatty acids with one or more carbon–carbon double bonds (C=C) are termed unsaturated and those with no such double bonds are saturated. The use of these terms is often extended to the fats and oils themselves, such that most people have come across the expression 'polyunsaturated fats' in relation to human diet.

Question 3.6

The selected dams show a greater *variation* for the characteristic being bred for than do the selected sires (i.e. the selected dams would have a greater standard deviation and variance than the selected sires). This can be seen from the relative lengths of the worst-to-best scale from which selected dams and selected sires have been drawn. Greater *numbers* of dams than of sires have been selected. This can be seen from the relative areas under the distribution curve in Figure 3.3a that represent selected dams and selected sires. Finally, the *mean performance* (or mean value of the characteristic being bred for) is higher (i.e. closer to 'best') for the selected sires than for the selected dams. Only the mean performance of *all* the parents is shown. However, bearing in mind the relative ranges in variation and numbers involved, the mean performance of the selected dams *must* lie a little to the left of that for all the selected parents and that of the selected sires must lie considerably to the right of it.

Question 3.7

It would be desirable in an experiment designed to estimate the heritability of a feature such as back-fat depth to treat the different generations *identically* (i.e. give them the same amounts of the same sorts of food, keep them in the same conditions and so on). Even if this were achieved, it would certainly be impossible to *prove* that they had been treated identically. The best that can be done is to arrange the experiment so that the effects on back-fat depth of any unplanned changes can be estimated. Since no selection is applied to the 'control' line, it can be used to estimate these effects so that they can be allowed for in analysing the responses of the selected lines. The main advantage of having a 'high' and a 'low' selection line is that *two* estimates of heritability can be made, such that their average is likely to be a more reliable estimate than either alone.

Question 3.8

Fat A is the more unsaturated; this can be seen from the data in the body of the table and more readily from the last line, which gives the ratios of the concentrations of unsaturated to saturated fatty acids. The value of 4.5 for fat A shows that it has more unsaturated than saturated fatty acids; the 0.6 for fat B shows that it has more saturated than unsaturated fatty acids.

Question 3.9

Comparing the ratios of unsaturated to saturated fatty acid concentrations in the carcass back-fat (1.3 for fat A in the diet and 0.8 for fat B in the diet), it is clear that the addition of the more unsaturated fat A resulted in a more unsaturated carcass back-fat than did the addition of the less unsaturated fat B.

Reasons for the ratios in the added fats not being the same as those in the carcass back-fat may include the fact that not *every* dietary fatty acid ends up in the carcass fat and the likelihood of there being sources of fatty acids other than fat A and fat B in the diet.

Question 4.1

Most of the nitrogen from the fertilizer (50–80%) ended up where it was intended, i.e. in the harvested parts of the crop (indeed, as protein in the actual *grain*).

The 10–25% that was in organic material in the soil would have been safe from leaching. (Although this nitrogen would gradually be mineralized, experiments of 4–5 years duration by the Rothamsted team have shown this to happen only very slowly.)

Fertilizer nitrogen left in the soil as nitrate at harvest is obviously vulnerable to leaching during the winter. Ammonium is also vulnerable because it is likely to be nitrified. The finding that only 1–2% of the ^{15}N was left in the soil as N_{min} by winter wheat is therefore of critical importance. It certainly casts doubt on the suggestion sometimes heard that the nitrate problem is caused by excess nitrogen fertilizer left unused by crops being washed away by winter rains.

The average loss of ^{15}N fertilizer between application and harvest was 15%. If the computer models were right, an average of only about 5% can have been lost by leaching (the rest having been lost by denitrification). Adding this to the 1–2% left vulnerable in the soil at harvest gives an estimate of 6–7% for the average amount of nitrogen fertilizer leached directly as nitrate. (Note that these figures relate to the winter wheat crop. Other crops, notably potatoes, leave more of the fertilizer in the soil as nitrate.)

Nitrogen fertilizer applied in autumn seems in most circumstances to be a waste of the farmers' money and an unnecessary load on the environment.

Question 4.2

(a) From the farmer's point of view, denitrification, like nitrate leaching, represents the loss of a resource. It is also not a process that can be controlled in the soil; in particular, there is no reliable way of ensuring that nitrogen (N_2), rather than nitrous oxide (N_2O), is produced. The best way of cutting the risk of nitrate leaching is to make sure that the crop takes up the nitrate before it can be leached from the soil rather than to rely on denitrification.

(b) Denitrification can be controlled far more precisely in a water-treatment plant than in the soil and conditions created which minimize the ratio of N_2O to N_2 released. Thus, denitrification is far more useful in this context. Of course, it is even better if the nitrate can be kept out of the water in the first place.

Question 5.1

The government might have been reluctant to make public findings about the possible link between BSE and CJD for the following reasons.

(a) The beef and related industries would be badly affected; incomes would drop and jobs could be lost.

(b) The government did not want to risk losing the votes of farmers and others employed in the beef industry.

(c) The government did not want to be seen to have failed to act early enough to protect consumers and to have told them that food was safe when it was not.

(d) Scientific evidence still provided no *direct* link between BSE and cases of CJD.

Question 6.1

Environmental testing of a new pesticide begins long before marketing and continues for a long time after the product appears on the market. While it may be worrying that the problem has been discovered *after* the pesticide has been made available to farmers (perhaps *many* years after its introduction), at least the pesticide's withdrawal demonstrates that the environmental testing system employed is effective.

Question 6.2

DDT is cheap and relatively safe to handle (it has probably never killed anybody using it on a farm). A poor farmer who cannot afford proper application equipment, and who lives in the tropics, where full protective clothing is uncomfortable if not dangerous to wear, cannot avoid contamination and inhaling the insecticide while applying it to the plants. Of the insecticides that can be afforded, the safest is probably DDT. On the other hand, exported along with DDT will presumably be some of its environmental problems, although the overall impact will be low in areas where less intensive forms of agriculture are practised.

Question 6.3

The long-lasting protection provided by organophosphate granules depends on a continual replacement of insecticide in the soil as the granules are degraded. Thus levels of insecticide in the soil will fall rapidly once the granule has finally dissolved completely. This is in marked contrast to insecticides in the highly persistent organochlorine group.

Question 6.4

A figure of 1.5% of spray landing on plants was quoted earlier. This would mean the effective amount of the 750 $\mathrm{g\,ha^{-1}}$ applied landing on the plants is 1.5% of 750 $\mathrm{g\,ha^{-1}}$, i.e. 11.25 $\mathrm{g\,ha^{-1}}$.

Question 6.5

Although setting 1 produces the more uniform spray, most of the drops it produces are too small to be deposited on plants. The majority have diameters of less than 80 μm, whereas only those with diameters of around 100–120 μm are likely to be deposited effectively. Thus, in this case, setting 2 is probably the better setting, although it produces the less uniform spray.

Question 6.6

The greater the proportion of a pest population killed by the pesticide, the greater the selection pressure in favour of resistant individuals. The greater the selection pressure, the more rapidly does resistance to the pesticide spread within the pest population and therefore the sooner the pesticide becomes ineffective. In the case of insecticides, such extremely toxic compounds are also likely to be damaging to natural enemies of both the pest in question and other *potential* pests (see later). There are occasions, therefore, when it is better to tolerate some loss in quality or yield in the short-term in order to maintain the pesticide's usefulness for longer.

Question 7.1

Barley growing in the ploughed soil had the greatest level of infection with BYDV and the lowest yield. Barley growing in soil with minimal cultivation had the lowest level of infection and the greatest yield. Barley growing in soil with reduced cultivation had intermediate levels of infection and yield. The incidence of disease was therefore lowest with the least intense method of cultivation. This was probably because the method of cultivation affected the survival rate of the predators of aphids which are disease carriers.

Question 7.2

Integrated farming techniques place emphasis on applications of agrochemicals to meet the needs of individual crops in order to reduce the amounts applied. Labour is needed to inspect for pests, diseases and weeds in the crop, for soil sampling prior to nutrient analysis and for mechanical weed control. Extra labour may also be required to harvest mixed crops grown in rows, strips or mosaics.

In order to harness natural control mechanisms, it is necessary to understand them. For pest control by natural predators, knowledge of the ecology of both predator and prey is required and for disease control, knowledge of economic thresholds can cut rates of application of fungicide. During inspections of their crops, farmers must also be able to identify, for example, signs of disease at an early stage.

Answers to activities

Activity 1.1

If there had been no significant changes in agricultural practice between 1936 and 1986, it might be expected that many of the agricultural statistics summarized in Tables 1.1 and 1.2 would result in ratios not too different from 1.2 (i.e. the ratio of the UK population in 1986 to that in 1936). Large deviations *above* 1.2 imply changes which cannot fully be accounted for by the need to provide for more people, while large deviations *below* 1.2 (particularly below 1.0) imply decreases relative to the increasing population size. It is not, however, possible to decide from the tables just how far a ratio has to deviate above or below 1.2 before the change can be regarded as *statistically* significant.

Parts (a) and (b) of Table 1.1, dealing with resources intrinsic to the farm and resources from outside the farm, contrast strikingly with one another. Perhaps surprisingly, the total area of land farmed in the UK declined over the 50 years (ratio <1.0). Since the number of farmers (taken to be equivalent to the number of farms) declined even more, farms became larger on average. The most dramatic reduction, however, is in the number of farm workers. This reduction is undoubtedly linked to the vast increase in the number of tractors (and presumably other farm machinery) during the same period (ratio >1.0). The dependence of modern farming on nitrogen (and presumably other) fertilizers produced away from the farm is also brought out clearly in the table.

Turning to part (a) of Table 1.2, greater amounts of all the grain crops except oats were produced in 1986 than in 1936. Indeed, an entirely new crop (oilseed rape) appeared during the intervening years. The reduced production of oats is presumably accounted for mainly by the demise of the farm horse, although human consumption of porridge oats has probably also declined! There is insufficient information to decide whether the increases in production are accounted for by increases in yield per hectare and/or by increases in the areas of land devoted to particular crops (it is clear from Table 1.1 that there has not been an overall increase in the area of agricultural land).

From part (b) of Table 1.2, it is clear that the production of potatoes per head of population hardly changed between 1936 and 1986. However, it is important to note that Table 1.2 relates to UK *production* rather than to UK *consumption*. For any of the commodities, the pattern of international trade could have changed dramatically over the 50-year period. The doubled per capita production of sugar beet probably reflects national policy to be less dependent upon imported cane sugar.

The switch from hay and various root crops to silage as feed for livestock is clear in part (c) of Table 1.2. In view of the comparatively modest increases in the production of beef and veal, mutton and lamb, liquid milk and wool evident in part (d), the increase in silage production must indicate quite significant changes in agricultural practice. The huge increases in the production of poultry meat, butter and cheese are probably mainly accounted for by profound changes in our eating habits, presumably related in part to greater affluence.

Activity 2.1

(a) We shall consider each of the three crops separately.

Wheat

(i) The area devoted to wheat increased fairly steadily from about 10^6 hectares to about 2×10^6 hectares over the two decades, i.e. it approximately doubled. (ii) During the same period UK production rose from about 4×10^6 tonnes to about 14×10^6 tonnes, i.e. it more than trebled. (iii) Yield (which is obtained by dividing production by crop area) increased from about 4 tonnes per hectare to about 7 tonnes per hectare from 1970 to 1990.

Barley

(i) Having remained fairly steady throughout the 1970s, the area devoted to barley declined steadily from about 2.2×10^6 hectares in 1980 to about 1.5×10^6 hectares in 1990. (ii) Production was 7.5×10^6 tonnes in 1970 and was only slightly greater, at about 8×10^6 tonnes, in 1990 (although it reached $10–11 \times 10^6$ tonnes in some of the intervening years). (iii) Yield rose from about 3.4 tonnes per hectare to about 5.3 tonnes per hectare over the two decades.

Oilseed rape

(i) The area devoted to oilseed rape has grown from about 15×10^3 hectares in 1973 to about 350×10^3 hectares in 1990, i.e. a more than 20-fold increase. The rate of increase was particularly great during the first half of the 1980s. (ii) From about 30×10^3 tonnes in 1973, production had grown to about 1000×10^3 tonnes in 1990, i.e. an increase of more than 30-fold. (iii) Yield therefore rose from about 2 tonnes per hectare to nearly 3 tonnes per hectare over this period.

(b) For all three crops, yield per unit area increased significantly over the period 1970–90. While weather may have affected yield in particular years, it is unlikely that climatic change over the 20-year period played a large part in this improvement. The two major components are likely to have been the use of new varieties (higher-yielding and more resistant to pests and diseases) and changes in farming practice (in particular, more effective use of fertilizers and pesticides).

Activity 2.2

What follows is *my* summary of the main points in the chapter.

1 Over the past 20 years oilseed rape has become an important feature of UK agriculture as a source of edible oil and protein feed for livestock.

2 Three of the four oilseed species in the genus *Brassica* arose as interspecific hybrids; doubling of the diploid chromosome numbers in the hybrids ensured that they were fully fertile.

3 Plant oils are triglycerides, i.e. esters of glycerol and three fatty acids.

4 Fatty acids with one or more carbon–carbon double bonds are termed unsaturated; those without any such double bonds are termed saturated.

5 Erucic acid was present at high concentrations in older varieties of oilseed rape; since erucic acid is potentially harmful, the low-erucic-acid character found in Canadian spring rape was bred into European winter rape. The initial reduction in yield, caused by the replacement fatty acids having fewer CH_2 units than erucic acid, was overcome by breeding for the production of more triglyceride molecules.

6 Different oilseed rape varieties have been bred according to the use to which their oil is to be put (e.g. margarine, cooking, engine lubrication).

7 Glucosinolates in rapeseed meal are broken down to give toxic products by the enzyme myrosinase released from crushed seed; this limited the use of rapeseed meal for animal feed despite its good balance of amino acids.

8 The genes responsible for low levels of glucosinolates in Polish spring rape were bred into high-yielding low-erucic-acid winter rape.

9 Most current oilseed rape varieties are double-low: low in erucic acid and low in glucosinolates.

10 Lower levels of glucosinolates may have implications for the interaction between oilseed rape and its insect pests.

It would be surprising if you chose precisely the same points to summarize as I have, let alone expressed them in the same way. The skill of summarizing is inevitably subjective, involving a compromise between comprehensiveness and brevity. If any of the above points are *completely different* from yours, check the importance of the point(s) in the text of the chapter.

Activity 3.1

(a) Our data for 24 eggs purchased in a supermarket are shown in Figure 3.5. With the exception of the one egg whose yolk corresponded to colour number 1 on the fan, the distribution of yolk colours is close to the classic so-called *normal distribution* typical of biological data (i.e. the frequencies fall away either side of a single most frequent colour—in this case, number 6). We examined too few 'free range' eggs obtained directly from a farm to draw reliable conclusions from these—but the eggs had noticeably darker yolks (colour number 9 being the most frequent) than the supermarket sample.

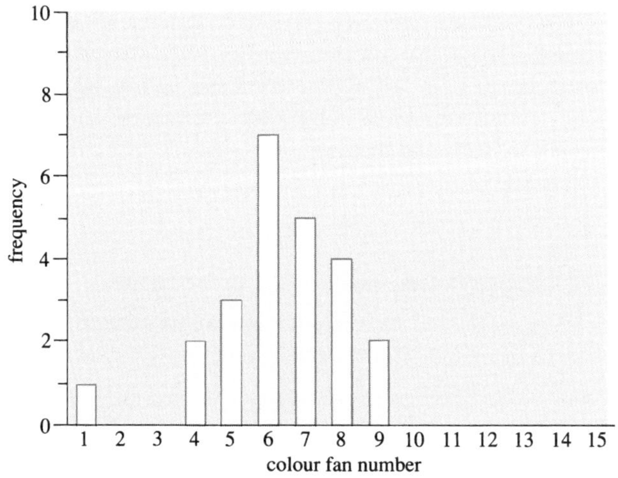

Figure 3.5 Yolk colours of 24 supermarket eggs.

(b) The preferences for egg yolk colour expressed by 47 adults are shown in Figure 3.6. Again, frequencies fall away either side of a single most frequent colour (in this case, number 7). However, this distribution is rather less like a normal distribution than the one we obtained in part (a). Possibly the most striking feature of the data is the wide range of preferences expressed by the subjects.

The data for the 29 women subjects (Figure 3.7) show essentially the same pattern as those for all 47 subjects. It is possible that the sample size of 18 is too small in the case of men for a clear pattern to be seen (Figure 3.8). There is no evidence in our data of a systematic difference in preferences for egg yolk colour between men and women. We did not investigate whether adults and children, or people with different ethnic or cultural backgrounds, differ in their preferences for egg yolk colour.

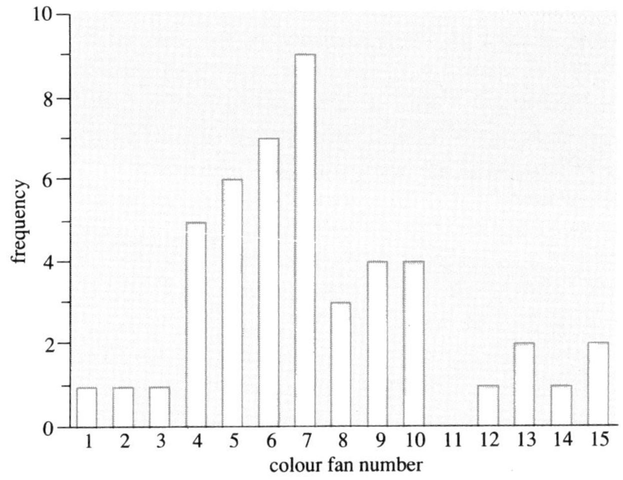

Figure 3.6 Preferred egg yolk colour of a sample of 47 adults.

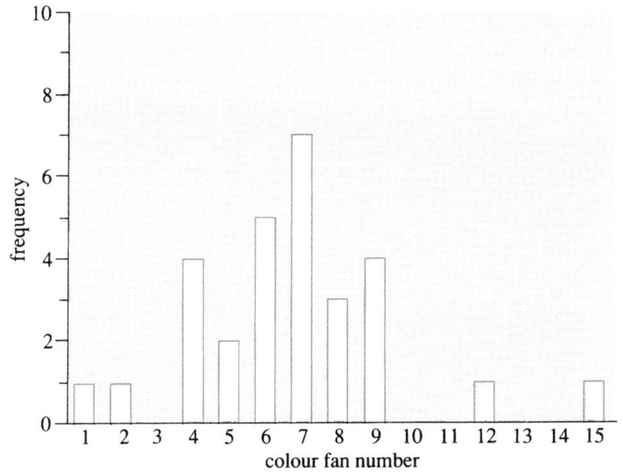

Figure 3.7 Preferred egg yolk colour of a sample of 29 women.

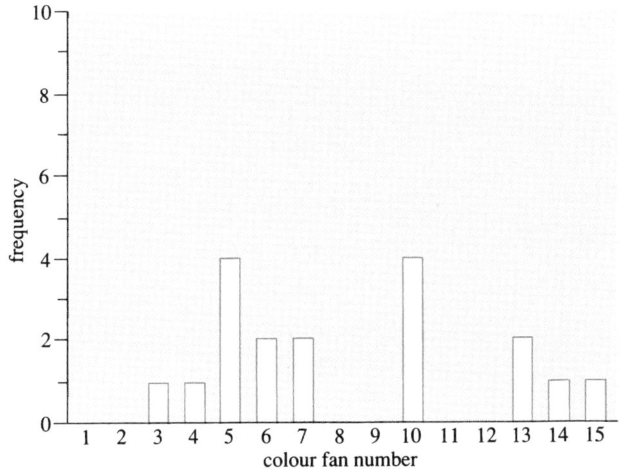

Figure 3.8 Preferred egg yolk colour of a sample of 18 men.

(c) Comparing Figures 3.5 and 3.6, we were impressed by the range of preferences for egg yolk colour expressed by our sample of people being *so much greater* than the range found in the 24 eggs purchased in a supermarket. Could it be that the major retail outlets, which have an increasingly powerful influence on the food we eat, dictate a somewhat arbitrary 'quality' criterion with respect to yolk colour which *must* be adopted by their suppliers? This would impose uniformity on the eggs sold—although that uniformity would have little to do with customer choice.

(d) We have asked you to compare egg yolk colours, and to elicit people's preferences for egg yolk colour, using a *colour photograph* of the yolk colour fan. It is quite possible that the colours of the fan have not been reproduced accurately in the photograph. In any case, you would certainly have found it easier to have used the fan itself; it is larger and its design would have allowed you to present the standard colours one at a time to the egg yolks or to have asked people to choose between two or three colours at a time (which may be better from a psychological point of view). Matching colours certainly involves an element of *subjective judgement.* In fact, 12 of our sample of 24 supermarket eggs were classified by one person (subject A) and the other 12 by another person (subject B). Overall, subject A judged her 12 eggs to be somewhat darker (Figure 3.9) than subject B judged *her* 12 to be (Figure 3.10). Although it is possible that the two sub-samples of eggs did differ from one another, we suspect there was a systematic bias on the part of one or both observers. It would have been better if we had asked the two people to have independently classified the *same* eggs.

How many eggs were you able to classify? Did you manage to collect preferences from a sample of 30 people? The larger the sample sizes, the more accurate conclusions based on them are likely to be. However, larger sample sizes are usually more time-consuming and costly to obtain. A 'trade-off' is therefore necessary. Optimum sample sizes depend on circumstances, but biologists often aim for at least 30.

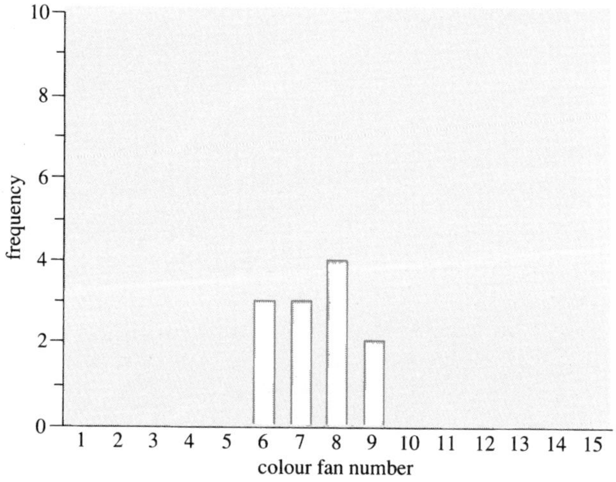

Figure 3.9 Yolk colours of 12 supermarket eggs classified by subject A.

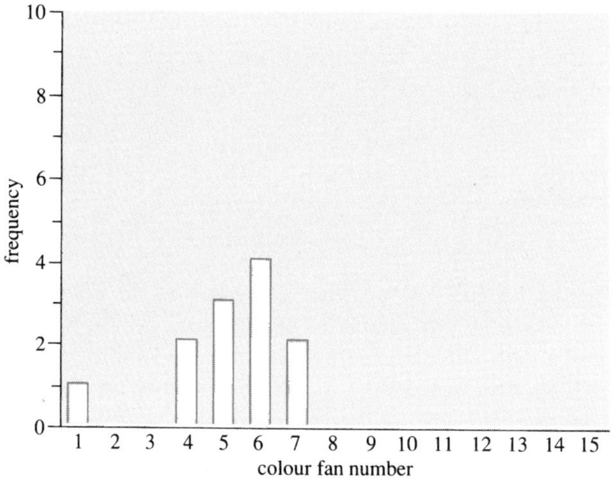

Figure 3.10 Yolk colours of 12 supermarket eggs classified by subject B.

Activity 4.1

Some *possible* relationships between the pairs of statements are given below.

(a) If statement 1 provides an explanation of statement 2, then perhaps Granny was right after all! An indirect relationship is rather more likely. The decline in the stork population and the decline in the human birth-rate are probably both related to a third factor, such as the increased size and affluence of the human population, and its more extensive and intensive use of land.

(b) Tempting though it might be to conclude that there were a lot of boozy vicars around at the turn of the century, an indirect relationship is again more likely. Both the need for clergy and the consumption of whisky increased as the population increased.

(c) At this stage you might well consider these statements to be directly related—indeed, *causally* related. That is, that in their search for profit, farmers applied increasingly large amounts of fertilizer—more than their crops could use—and that the excess was washed into rivers and other natural waters. Farming *has* certainly undergone massive changes in the past century and some of these changes *have* led to increased leakage of nitrate from farmland. However, the use of more fertilizer is only *one* of these changes and increased leakage of nitrate is *not* simply accounted for by excess fertilizer being washed away. (This may seem not so much an indirect relationship, as an obscure one! You are therefore unlikely to have suggested it. By the time you have studied this chapter you should see what we mean.)

Activity 4.2

Here is one person's view on the risk of methaemoglobinaemia.

Methaemoglobinaemia is a terrible condition and simply must not be allowed to happen. Fortunately it hardly ever does in this country, presumably because baby feeds are almost always made up with tap-water. (I am not aware of any cases caused by the feed being made up with tap-water.) Thus, I do not believe that there is any need for widespread worry about the condition. We should, however, remain alert to the possibility that nitrate concentrations in (say) the 50–100 $mg\,l^{-1}$ range could cause subclinical cases of methaemoglobinaemia.

If nitrate concentrations in a water supply do become undesirably high, baby feeds can be made up with bottled low-nitrate water. But, a cautionary word: although the 'mineral water' that you can buy in supermarkets usually contains much less nitrate than tap-water (the label tells you how much), it often contains more bacteria than tap-water. The Consumers Association recommends that bottled water should be boiled before it is used in baby feeds.

The European Commission is a cautious body and its nitrate limit seems to have a built-in safety factor of at least 2. That is, the limit is half the minimum concentration likely to cause problems.

Making up baby feeds with well-water of poor quality led to the last recorded death from blue-baby syndrome in the UK, a death that would not have occurred if the baby had been breast-fed rather than bottle-fed. It is my personal opinion that mothers in developing countries should not be persuaded by advertizing, as they sometimes are, to use powdered baby-milk when they have plenty of breast-milk but water of very poor quality.

Activity 4.3

I take cancer seriously (in fact, I almost died of a brain tumour). But I am equally sure that the last thing we should do is panic about it. Fear is the worst possible basis for taking decisions.

- Consider the trend in nitrate concentrations in water (which have increased generally during the last 30 years; see Figure 4.2) and the trend in incidence of stomach cancer (which has declined). Someone must produce some very strong evidence for a link between nitrate in water and stomach cancer before we can dismiss the fact that these trends go in *opposite directions*.
- Epidemiologists have put a lot of effort into looking for the kind of evidence we are talking about. The saliva test survey (Table 4.1) provides no evidence for such a link. In the more reliable refined samples, there are no statistically significant differences (i.e. $P>0.05$) between the mean nitrate concentrations in saliva for the two low incidence areas (I and II) or for the two high incidence areas (III and IV). Although the difference is highly significant (i.e. $P<0.0001$) when the two low incidence areas and the two high incidence areas are combined, the mean nitrate concentration in saliva samples is *lower* in the areas with a high incidence of stomach cancer. I also know of another study (not mentioned here) that looked at the relationship between stomach cancer in urban populations and nitrate in their water supplies and found equally little evidence for the supposed link.
- And the workers in the fertilizer plant (Table 4.2)? Well, all I can say is that, from a health point of view, I should be more than happy to work in a fertilizer plant. For stomach cancer there are no significant differences (i.e. $P>0.05$) between observed and expected mortalities for heavily exposed workers, for less heavily exposed workers and for both groups of workers together. In fact, observed mortality from all causes is *lower* than that expected for heavily exposed workers (though not significantly), for less heavily exposed workers ($P<0.01$) and for both groups of workers together ($P<0.001$).

What possible biases and hidden factors did you find in the data?

- In the saliva survey, could either the incidence of stomach cancer or the nitrate concentrations in saliva have been influenced by another factor—for example, the eating habits of the people in the four areas? For instance, vitamin C is suspected to be an anti-cancer agent. Thus, if the people in the low-incidence areas ate more fruit and vegetables than those in the other areas this would have biased the study. In fact, the epidemiologists considered this possibility and took it into account in their statistical calculations—and still found no evidence that stomach cancer was related to nitrate exposure.
- Weren't the fertilizer workers almost suspiciously healthy? It wasn't just that they had no more stomach cancer than the other workers; they had less respiratory and heart diseases as well (Table 4.2; though the differences were statistically significant only for both groups of workers together). They worked for a company that looks after its workers well, so they might have been generally healthier and less stressed. However, I would go for the 'big bang' theory: ammonium nitrate and dust explode readily when mixed; my guess is that the workers in the fertilizer plant were strongly discouraged from smoking, and this would account for a lot.
- Cancer does not develop instantaneously. There is inevitably a lag between the event that causes the cancer and its development to the extent that the patient feels ill enough to go to a doctor. This lag may be of years or even tens of years. Therefore, how relevant is it to relate stomach cancer *now* to exposure to nitrate *now*? Unfortunately, there are insufficient data on levels of nitrate in water 20, 30 or 40 years ago and the limited information that we do have cannot be evaluated unequivocally.

- Nitrate, and also nitrite, are present in the preserved meats and vegetables that we eat. Shouldn't we take these into account? Since only about half the nitrate we consume comes from water, we certainly should. A Canadian study that did look in detail at this showed no association between cancer risk and nitrate in food, but a definite association between cancer risk and nitrite in food.

My own opinion is that there is no evidence to link stomach cancer to nitrate in our water supplies. This is in line with established medical opinion; the government's Chief Medical Officer issued a statement in 1985 accepting that there was no evidence for such a link.

Activity 4.4

Ammonification The production of ammonium from organic nitrogen.

Biomass The total amount of living organic matter (in the soil in this case), i.e. both microbes and larger organisms.

Decomposition The process by which organic matter in the soil is gradually broken down first by larger organisms and then by microbes to form organic nitrogen, which is ultimately mineralized.

Eutrophication The proliferation of algae caused by enrichment of water with nitrate and/or phosphate, leading to the death of (for example) fish as a consequence of the oxygen supply in the water being exhausted by algae-decomposing bacteria.

Fixation The conversion of atmospheric nitrogen to ammonium by either specialized algae and bacteria or the chemical industry, as a stage in the production of nitrate; the direct production of oxides of nitrogen from atmospheric nitrogen by lightning.

Humus Dead organic material in the soil, mainly derived from plants.

Leaching The process by which nitrate is washed beyond the reach of plant roots in the soil and hence ultimately into rivers, lakes and/or aquifers.

Mineralization The production of nitrate and ammonium (mineral N or N_{min}) from organic nitrogen in the soil, i.e. ammonification plus nitrification.

Nitrification The conversion by microbes of ammonium to nitrate via nitrite.

N_{min} Mineral N (i.e. nitrate and ammonium) in the soil.

Activity 4.5

My list of the important (i.e. key) points so far is as follows.

1 The concentration of nitrate in water in rivers, lakes and aquifers has been increasing at the same time as the amount of nitrogen fertilizer used in agriculture has increased.

2 Nitrate is basically an environmental rather than a health problem.

3 In freshwater, excess nitrate usually causes an environmental problem (eutrophication) only in the presence of excess phosphate (which may come from sewage).

4 Nitrogen is fixed from the air by algae and bacteria as well as in the manufacturing of fertilizers.

5 Nitrate is added to the soil by soil microbes breaking down organic matter, as well as by the farmer applying fertilizer.

6 Farming based on arable crops and farming based on animal husbandry are *both* liable to lead to leaching of nitrate.

7 All nitrate is the same, regardless of origin.

8 The root of the nitrate problem lies in the 'equation':

availability = vulnerability

i.e. if nitrogen in the soil is *available* to plants as nitrate, it is by the same token *vulnerable* to leaching.

9 We are looking at not so much a nitrate problem as a problem of *untimely nitrate*.

Your list is bound to be somewhat different from mine. You will undoubtedly have *expressed* points differently. However, if any of your points are *completely different* check whether they are *directly relevant* to the nitrate problem and, if they are, whether they can really be described as *important*.

Activity 4.6

(a) Nearly all the nitrogen fertilizer given to winter wheat is applied in spring. Farmers used to give some nitrogen to winter wheat in autumn to 'help it through the winter', but this is done to a much lesser extent now.) In most of the Rothamsted experiments, therefore, the tagged nitrogen was applied in April, although in one or two experiments it was applied in autumn.

(b) Although the amount of fertilizer applied was varied experimentally, it was obviously important to apply an appropriate amount of fertilizer and to apply it evenly. You might also have thought of applying the tagged nitrogen to a relatively small area in the centre of the plot, since plants around the edge are likely to grow (and therefore use fertilizer) differently from those surrounded by other plants. (You would not, incidentally, need to take precautions against radiation hazards because ^{15}N is not a radioactive isotope.)

(c) In each plot, the Rothamsted team looked for ^{15}N in:

the grain

the straw

the roots and stubble

the chaff (in some experiments)

the organic matter in the soil

the N_{min} in the soil, i.e. the ammonium and nitrate. (This was very important because it is this nitrogen that is vulnerable to leaching.)

weeds growing in the plot

(d) It is unreasonable to expect to recover all the ^{15}N. Some tagged nitrogen would be lost downwards into groundwater (i.e. leached). Some would also be lost upwards as gases (see later). Comprehensive measurements of such losses are very difficult and were not feasible in these particular experiments.

Activity 4.7

Either applying no fertilizer or severely limiting the amount applied (in particular, not applying any in autumn when little use can be made of it by crops) helps to reduce the amount of nitrate vulnerable to leaching from agricultural land. Not leaving land bare in the winter (in other words, growing a winter or a 'catch' crop) minimizes leaching of nitrate released through microbial action in the soil in autumn. The ploughing of grassland (other than that in rotation) is limited because this practice releases enormous amounts of nitrate from organic matter in the soil, much of which is then leached. Apart from their other advantages, woodlands (and presumably hedgerows) are more 'closed' with respect to nitrogen-cycling than either arable or pastoral farmland and are therefore much less likely to contribute to the nitrate problem. Intensive animal production units could contribute greatly to the nitrate problem if adequate arrangements are not made for dealing with the large amount of manure produced in a relatively small area.

Activity 4.8

Does your personal strategy for combating the nitrate problem include any of the following?

Putting nitrate into the environment

- Did you realize that you excrete 6 kg of nitrogen into the environment each year in forms that are readily converted to nitrate? There is not much you can do about this, but you can check that your water company is treating sewage properly.
- If you are a gardener, do you use excessive amounts of artificial fertilizer or manure? It is much better to recycle garden waste by composting.

Washing nitrate into the environment

- When you water your garden, do you water plants individually or do you water large areas of bare soil as well? Do you have any idea how much water you apply? Excessive watering washes nitrate downwards just as effectively as rain.
- Surely you don't put fertilizer or manure on to your garden and then wash the nitrate out with a sprinkler!

Taking water out of the environment

Nitrate concentrations in water are increased by having more nitrate in the system—or less water. The more water that is extracted from rivers and aquifers, the less there is to dilute any nitrate that gets into them. So water should be used sparingly.

- Bathe with the environment in mind. How about having one less inch of water in the bath? You might not want to share a bath, but what about sharing the bath water? Or get the children to share it.
- Flush with discretion. Even if you can't afford a low-volume cistern, you can still save one litre per flush. Wash two 500 g plastic margarine containers and their lids. Sink the containers in corners of the cistern where they do not obstruct the works. Put on the lids. At a stroke you have immobilized one litre of water!
- Do you have a dishwasher? Do you know how much water it uses on each run? Do you use it only when it is full up? Why not enhance family life by washing up together occasionally?
- Watering the garden? Take a long hard look at that sprinkler!
- Car washing? With bucket and sponge—or using a hose with water running away down the street? An automatic car-wash? Really!
- Are there managerial or technical improvements that will help your work-place to use less water?

Cleopatra's critics considered her an extravagant hussy for bathing in asses' milk. What would those critics make of people who spend millions of pounds getting nitrate out of water, so that the water is fit to be added to baby's powdered milk, and then bathe in the water, flush the loo with it, wash the car with it and sprinkle it on the garden—so that more nitrate dissolves in it?

Causing others to put nitrate into the environment

- Do you eat more meat than you need to? Meat-producing animals do not convert nitrogen efficiently.
- Could any of your investments be squeezing a farmer to use more nitrogen than is really efficient?
- Have you invested in a water company? If so, are you keeping a sharp eye on its policies?

Activity 5.1

(a)(i) At first glance you may have thought that Figure 5.3 shows the percentage of animals with BSE in each age group. However, inspection of the numbers on the *y*-axis shows that the total does not add up to 100%. In fact, 'age-specific incidence in BSE-affected herds' means, for each age group, the number of affected animals as a percentage of the total number of animals in that age group *in the affected herds*. Note that the data do not relate to the national herd as a whole. When asked to interpret presented data, it is very important to think carefully about the precise meaning of any term used.

(ii) Superficially, it can be seen that: the youngest animals affected are 2 years old; incidence peaks at 4 years of age and it declines thereafter. With no further knowledge of cattle production one would be left trying to work out what was special about cattle that were 4 years old.

(iii) Cattle start to breed at 2 years old. Therefore BSE may be related to breeding. On the other hand, TSE diseases have long incubation periods and so at least some of the younger infected cattle may have become infected before reaching reproductive maturity.

(b)(i) Few male cattle are kept for breeding because of the widespread use of artificial insemination. Male cattle are raised for beef but they are killed between 1 and 2 years old. Dairy herds and beef breeding herds consist of female cattle, so it is not surprising that more females than males were affected by BSE. In fact, in England and Wales in 1987, there were 3 200 000 female and 37 000 male cattle.

(ii) The incidence in male cattle is 2 in 37 000; expressed as a percentage this is:

$$\frac{2}{37\,000} \times 100 = 0.005\,4\%$$

The incidence in females is 190 in 3 200 000; expressed as a percentage this is:

$$\frac{190}{3\,200\,000} \times 100 = 0.005\,9\%$$

Therefore the percentage incidence is similar in the two sexes.

(c)(i) The incidence of BSE in dairy cows is $\frac{696}{2\,324\,000} \times 100 = 0.030\%$.

(ii) The incidence of BSE in dairy cows is about 15 times greater than in beef suckler cows.

(d)(i) The main source of BSE cases in dairy herds was homebred cattle but in beef herds it was purchased cattle.

(ii) Because excess calves from the dairy herd are sold and reared as beef cattle, it is possible that most of the beef cattle with BSE came from dairy farms.

(e)(i) If number of affected herds was plotted, counties/regions with many dairy herds (e.g. Cheshire and Somerset) would be likely to have a higher incidence of BSE just because they had more herds. Plotting the number of herds affected as a percentage of the total number of herds in each county/region makes some allowance for this and hence interpretation of the map easier. Of course, the authors could have chosen to plot the data in other ways, e.g. the number of affected dairy cows in each county/region as a percentage of the total number of dairy cows in that county/region.

(ii) No particularly clear patterns are obvious. However, counties in the Midlands and in the south east of England generally had a higher incidence of BSE than elsewhere, the highest values (4% and greater) being found in Kent, Hampshire, Berkshire and Bedfordshire. Notable anomalies were Cornwall with a rather high, and Essex with a rather low, incidence. Borders and Lothian also had a noticeably higher incidence than the adjacent counties/regions. The zero percentage incidence in London and Hertfordshire may be a reflection of the low numbers of dairy herds in these areas, but those throughout Scotland must surely be significant.

Activity 5.2

(a) Southwood said that a ban on cattle brains from the start (presumably from June 1988 when the Southwood committee was set up) would have stood no chance of acceptance by MAFF. He pressed for a ban which was implemented in November 1989. The Ministry wanted a one year ban but Southwood insisted on a permanent ban. Southwood advised full compensation for infected animals that were destroyed, but MAFF would not agree to this at first. 50% compensation was given from August 1988 until February 1990, after which full compensation was awarded. It seems that the Ministry eventually followed the recommendations of the scientists, but only after some time. When the proposals were first suggested, MAFF considered many of them to be revolutionary.

(b) Evidence from scrapie: humans have eaten meat infected with the scrapie agent for many decades without harm. The consequence of the view that the 'risks to humans were remote' was that some farmers carried on using contaminated feed for many months and evaded bans on selling diseased animals. Abattoirs bent the rules on the separation of SBOs from meat for human consumption and the authorities turned a blind eye to the non-compliances.

(c) Some farmers passed off sick animals as healthy because they were given only 50% compensation (from August 1988 to February 1990) for the slaughter of cattle with BSE. Although not discussed in the extract, farmers can often recognise changes in behaviour in their cattle before the appearance of classic symptoms of BSE. Such animals could be quickly sold at market and the farmer not penalised for having a BSE-infected beast.

Activity 6.1

(i) In order to draw valid conclusions from the results of such an experiment, it is important that the plants used are *representative* of the species or variety. Experimental plants *may* differ from one another genetically; but they will almost certainly vary as a result of growing in slightly different environments. It would therefore be unwise to base conclusions on only one plant per treatment (as can be seen from the variation in number of dead aphids *between* replicate plants *within* treatments). On the other hand, given the intricate nature of this experiment, it would be very time-consuming (and therefore expensive) to use hundreds of plants per treatment. The use of six *replicate* plants per treatment is therefore a compromise.

(ii) It is to be expected that some aphids will die as a result of natural causes and that some will die as a result of handling, irrespective of whether or not they are exposed to the insecticide. There is little point comparing the mortalities associated with two insecticide treatments if neither is significantly above this 'baseline' level. Treatment 1 is the *control* for this experiment.

(iii) In treatment 2, the insecticide can reach the aphids only if it is transported from the soil to the leaves of the plant; the route is therefore *systemic* (presumably involving the water-transporting *xylem*). Treatment 3 tests the effect of direct *contact* between the aphids and the insecticide without the latter possibly being routed through the plant. Treatment 4 is the *residual* route. Since the insecticide is painted on the top side of the leaf in treatment 5 and the aphids are caged on the underside, the route is *translaminar*. Treatments 6 and 7 both involve *systemic* transport of insecticide within the leaf; the tip-to-base direction in treatment 6 suggests transport in the *phloem*, while the base-to-tip direction in treatment 7 suggests transport in the *xylem*. The only route to target available in treatment 8 is the *fumigant* one. In fact, the fumigant route to target might possibly also apply in an uncontrolled way to several of the other treatments (particularly 4–7, which involve painting insecticide onto the leaves).

(iv) Totalling shows that the aphid mortalities in the different treatments (i.e. routes to target) were as shown in Table 6.5.

(v) The only route to target that failed to give mortality much above the level of control treatment 1 was systemic (phloem) (i.e. treatment 6). The insecticide clearly shows systemic (xylem) (treatments 2 and 7), contact (treatment 3), residual (treatment 4) and fumigant (treatment 8) properties. It is probable that the insecticide also shows translaminar (treatment 5) properties. As mentioned above, the fumigant route to target (which resulted in a high percentage mortality in treatment 8) could possibly complicate interpretation of the results for treatments 4–7; however, the low percentage mortality in treatment 6 suggests that this effect is not particularly important.

Table 6.5 Aphid mortalities in different treatments.

Treatment	Route to target	No. dead (out of 60)	Mortality/%
1	control	11	18
2	systemic (xylem)	60	100
3	contact	60	100
4	residual	60	100
5	translaminar	46	77
6	systemic (phloem)	14	23
7	systemic (xylem)	51	85
8	fumigant	57	95

Activity 6.2

(a) Approximately 200 drops m^{-2}.

(i) The diameter of each drop is 1 mm, therefore the radius (r) is 0.5 mm. Using the formula $V = \frac{4}{3}\pi r^3$, the volume (V) of each drop is thus $\frac{4}{3}\pi(0.5)^3$ or $0.524\,mm^3$.

(ii) 1 litre is equivalent to $1\,000\,cm^3$. A litre can therefore be pictured as a cube with 10 cm sides ($10\,cm \times 10\,cm \times 10\,cm = 1\,000\,cm^3$). Since 10 cm is equivalent to 100 mm, the volume of the cube is also $100\,mm \times 100\,mm \times 100\,mm$ or $10^6\,mm^3$. Thus 1 litre is equivalent to $10^6\,mm^3$.

(iii) The number of identical 1 mm diameter drops that can be obtained from 1 litre of spray is given by dividing the total volume ($10^6\,mm^3$) by the volume of each drop ($0.524\,mm^3$). The answer is 1.91×10^6.

(iv) There are $10^4\,m^2$ in a hectare (i.e. $100\,m \times 100\,m$).

(v) The number of identical 1 mm diameter drops in each $1\,m^2$ is given by dividing the total number of drops (1.91×10^6) by the number of $1\,m^2$ there are in 1 hectare (10^4). The answer is $1.91 \times 10^2\,m^{-2}$, which is approximately 200 drops m^{-2}.

If you have not yet attempted to answer part (b), do not read on until you have.

(b) Approximately 2×10^8 drops m^{-2}. Reducing the diameter of the drops to one hundredth of their former size, increases the number of drops by one million!

You can answer this question in essentially the same way as the first. The volume of each drop is $\frac{4}{3}\pi(0.005)^3$ or $5.24 \times 10^{-7}\,mm^3$. The number of identical 0.01 mm diameter drops is given by dividing the total volume ($10^6\,mm^3$) by the volume of each drop (5.24×10^{-7} mm^3), and is thus 1.91×10^{12}. The number of identical 0.01 mm diameter drops in each 1 m^2 is given by dividing the total number of drops (1.91×10^{12}) by the number of 1 m^2 in 1 hectare (10^4), and is thus 1.91×10^8 drops m^{-2} (or approximately 2×10^8 drops m^{-2}).

A quicker and more direct way is to appreciate that the volume of a sphere depends on the *cube* of its radius. If the radius of the drops is 100 times smaller, the volume of the drops must be 100^3 (or 10^6) times smaller and it should therefore be possible to obtain 10^6 times the number of drops from the same volume of spray.

Activity 6.3

(i) The accumulated data for setting 2 are shown in Table 6.6.

Table 6.6 Accumulated data for setting 2 (Table 6.2b).

Setting 2 (flow rate: 1.0 litre min^{-1}; rotation speed: 2000 rev. min^{-1})		
Drop diameter/μm	No. of drops in diameter range	% volume in diameter range
0–50	12	0.01
0–100	72	0.97
0–150	176	8.68
0–200	280	29.80
0–250	360	64.35
0–300	392	89.58
0–350	400	100.00

(ii) Graphs of the accumulated numbers of drops and the accumulated volumes for settings 1 and 2 are shown in Figures 6.4 and 6.5 respectively.

(iii) It depends on how you draw your curves, but the answers you obtain should be fairly close to those in Table 6.7 (which are expressed to the nearest 5 μm).

Table 6.7 Mean drop diameters for settings 1 and 2.

Setting	Number mean diameter/μm	Volume mean diameter/μm
1	60	70
2	160	230

(iv) Using the data in Table 6.7, the volume mean diameter/number mean diameter ratios are 70/60 = 1.2 for setting 1 and 230/160 = 1.4 for setting 2. Thus setting 1 produces the more uniform spray.

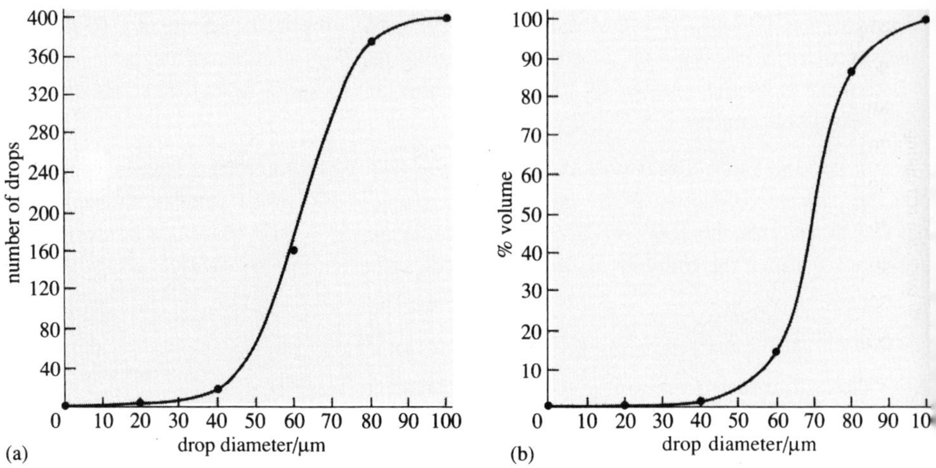

Figure 6.4 Cumulative plots for setting 1. (a) Number of drops against drop diameter. (b) % volume against drop diameter.

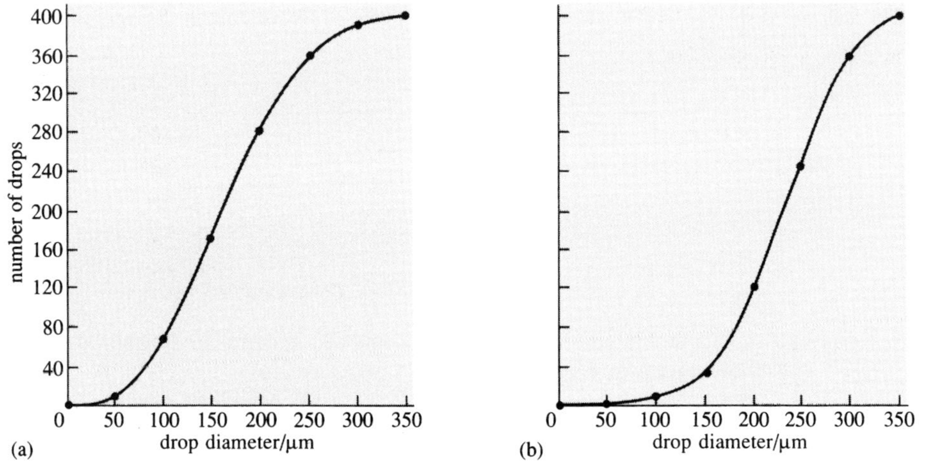

Figure 6.5 Cumulative plots for setting 2. (a) Number of drops against drop diameter. (b) % volume against drop diameter.

Activity 6.4

(a) The graph is shown in Figure 6.6a. The pattern is of an exponential or logarithmic decay.

(b) Reading across from 50% of the pesticide remaining to the curve and then down to the x-axis, the half-life is seen to be approximately 8 days.

(c) There are two ways to answer this question: arithmetically from the half-life derived in part (b) and graphically from the exponential decay curve plotted in part (a).

If 50% remains after about 8 days, then half of *this* (i.e. 25%) will remain after about another 8 (=16) days, and so on. The amount remaining will therefore be 12.5% after about 24 days, 6.25% after about 32 days, 3.125% after about 40 days and 1.5625% after about 48 days. Thus less than 3% will remain after something over 40 days.

One of the problems about using exponential decay curves, either to estimate half-lives (part (b)) which are then used in subsequent calculations or to estimate when less than a particular percentage of the original will remain (as here), is that it depends on just how you draw the curve. Extrapolating beyond 32 days in Figure 6.6a, we estimate that less than 3% will remain after about 42 days.

(d) The graph for $\log_{10}$ of percentage of pesticide remaining is shown in Figure 6.6b. It is plotted from the data in Table 6.8.

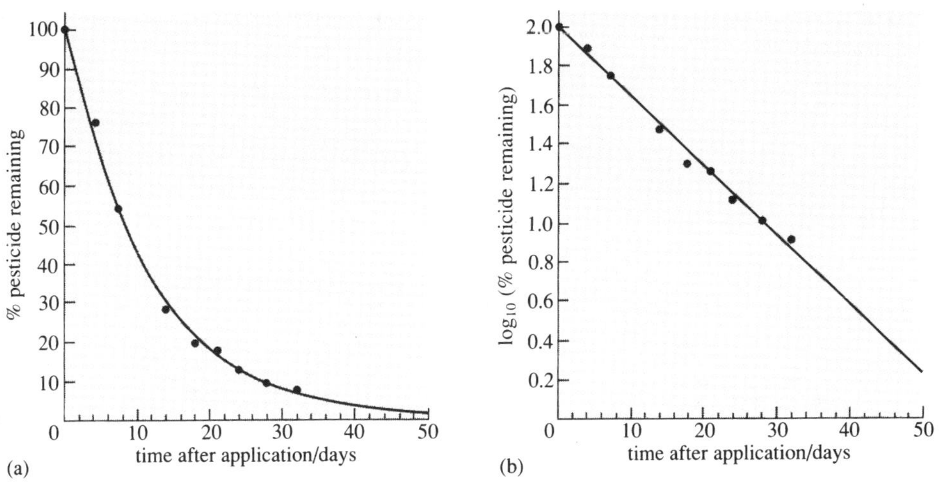

Figure 6.6 (a) Percentage of pesticide remaining on leaf surface against number of days since application. (b) Log_{10} of percentage of pesticide remaining on leaf surface against number of days since application.

Table 6.8 Data for Figure 6.6b.

Time after application/days	% pesticide remaining	Log_{10} (% pesticide remaining)
0	100	2.00
4	76	1.88
7	54	1.73
14	29	1.46
18	20	1.30
21	18	1.26
24	13	1.11
28	10	1.00
32	8	0.90

A straight line can be fitted to the $\log_{10}$ plot, confirming that percentage of pesticide remaining does indeed decay exponentially. It is *much* easier to fit a straight line to a set of points than to draw a smooth curve through them. The half-life subsequently read off a straight-line graph is thus more likely to be accurate than one read off a curve. Noting that $\log_{10}(50) = 1.70$, you should again have found that the half-life is approximately 8 days. It would also be much easier to determine how long it would take until less than 3% of the pesticide remained by extending the straight line and reading across from 0.48 (since $\log_{10}(3) = 0.48$). Again, the answer we get is approximately 42 days.

Activity 6.5

(a) Figure 6.3 shows that resistance to the pesticide is distributed *normally* in both generations (i.e. as a bell-shaped curve centred on the mean level of resistance, the LD_{50}). If the LD_{50} dose were applied, then all those individuals to the left of the dashed lines would be killed while all those to the right would survive (i.e. are resistant to that dose). In fact, the dose applied to each generation (X) was higher than the LD_{50} and *all* those individuals in the unshaded parts of the distributions were killed. The second generation was therefore bred entirely from individuals who were resistant to levels of the pesticide greater than X. The second generation displays rather less variation in resistance to the pesticide than does the first (i.e. the peak is narrower and taller). The dose of pesticide corresponding to LD_{50} has increased. Finally, a greater proportion of the second generation than of the first possesses the ability to survive a dose X of the pesticide (compare the relative proportions of the shaded parts of the distributions).

(b) If resistance was to some extent genetically determined, then it would be expected that the offspring of survivors would be somewhat more resistant to the pesticide than was the first generation *as a whole*. That the LD_{50} *is* higher after selection than before shows that resistance was indeed (at least partially) genetically determined. Thus, the operation of natural selection has increased the level of resistance to the pesticide in the pest population and would be expected to carry on doing so until the genetically-based variability in resistance to the pesticide within the population has been exhausted.

Activity 6.6

Any order is bound to be subjective, but experience over the 50 years of large-scale pesticide use would suggest the following (in order of decreasing importance):

resistance to pesticides

destruction of natural enemies

effects on larger wildlife

environmental contamination

hazards to humans.

Activity 7.1

Your list may have included some of the following (and perhaps some we have not though of):

1 Overproduction and accumulation of food in EU food 'mountains' in costly stores.

2 Eutrophication of water caused by over use of nitrogen fertilizer, and contamination with livestock waste and silage effluent.

3 Possible human health problems from contamination of water and food with pesticide residues and nitrates.

4 A decrease in diversity of natural plants and animals resulting from over use of herbicides and nitrogen fertilizer.

5 Flooding and loss of soil from large fields after high rainfall.

6 A decrease in jobs and an increase in rural poverty with the increase in farm size and greater mechanisation.

7 A loss of character in the countryside with the removal of hedgerows and woodlands, drainage of wet areas and specialisation of farms.

8 Inhumane treatment of farm animals.

9 Loss of taste in farm produce.

10 The overprotection of European farmers causing problems in talks on international trade reform.

Activity 7.2

(a) The average yield for wheat in the UK in the early 1990s was about $7\,t\,ha^{-1}$, so the yield from the conventional system of $8.2\,t\,ha^{-1}$ is considerably greater and the yield from the integrated system of about $7.0\,t\,ha^{-1}$ is equal to the UK average.

(b)(i) In all cases, the yields of crops in the integrated system are lower than the yields of crops in the conventional system. (ii) The standard deviations of yields of oilseed rape and barley/oats are about the same in each system. The standard deviations of yield for wheat in the integrated system are about half of those for the conventional system. So for wheat at least, yields in the integrated system appear to be more stable.

(c) See completed Table 7.6.

Table 7.6 Yields, costs of inputs and gross margins of the integrated system as a percentage of those from the conventional system in the LIFE project.

	Yield/%	Cost of inputs/%	Gross margin/%
winter wheat 1	84	57	94
winter wheat 2	87	64	117
winter oats/barley	81	45	113
winter/spring oilseed rape	74	62	102

(d) Gross margins were similar or better (102% to 117%) in the integrated system for three of the crops (winter wheat 2, oats/barley and oilseed rape) and only a little less for winter wheat 1 (94%).

For a gross margin of 100% or more, the loss in revenue from lower yields (74% to 87%) must have been compensated for by lower costs of inputs (45% to 64%).

Note: In the case of winter wheat 1, the integrated rotation relied upon a disease resistant variety, 'Pastiche', which was lower yielding than the variety used in the conventional rotation.

(e) Incorporation of a nitrogen-fixing legume into the rotation might increase the nitrogen content of the soil. It would be important that the legume does not result in the build up of nitrate in the soil at times of the year when there was a risk of leaching.

Acknowledgements

The Course Team would like to acknowledge the help and advice of the external assessor for this book, Professor Dennis A. Baker of Wye College, University of London.

Grateful acknowledgement is also made to the following sources for permission to reproduce material in this book:

Text

Extract 5.1 Pearce, F. (1996) 'Ministers hostile to advice on BSE', *New Scientist,* 30 March 1996, IPC Magazines.

Figures

Figures 4.1, 4.4 and 4.5 Addiscott, T. M., Whitmore, A. P. and Powlson, D. S. (1991) *Farming, Fertilizers and the Nitrate Problem*, CAB International; *Figure 4.2* adapted from Department of Environment (1988), *The Digest of Environmental Protection and Water Statistics*, No. 11, reproduced with the permission of the Controller of Her Majesty's Stationery Office; *Figure 4.6* adapted from Chaney, K. (1990) 'Effect of nitrogen fertilizer rate on soil nitrogen content after harvesting winter wheat', *Journal of Agricultural Science*, **114**, pp. 171–176, Cambridge University Press; *Figures 5.1 and 5.8* MAFF (1996) 'Bovine spongiform encephalopathy in Great Britain: A progress report', May 1996, © Crown Copyright. Reproduced with the permission of the Controller of Her Majesty's Stationery Office; *Figure 5.2* Crown Copyright, Veterinary Laboratories Agency, 1996; *Figures 5.3 and 5.6* Wilesmith, J. W., Ryan, J. B. M. and Atkinson, M. J. (1991) 'Bovine spongiform encephalopathy: epidemiological studies on the origin', *The Veterinary Record*, **128**, pp. 199–203, British Veterinary Association; *Figure 5.4* Bradley, R. and Wilesmith, J. W. (1993) 'Epidemiology and control of bovine spongiform encephalopathy (BSE)', *British Medical Bulletin,* **49**(4), 1993, The British Council; *Figure 5.5* Wilesmith, J. W., Wells, G. A. H., Cranwell, M. P. and Ryan, J. B. M. (1988) 'Bovine spongiform encephalopathy: epidemiological studies', *The Veterinary Record*, **123**, pp. 638–644, British Veterinary Association; *Figure 6.1* Micronair, Sandown, Isle of Wight; *Figure 6.2* Micron Sprayers Ltd, Bromyard; *Figures 7.1 and 7.2* Boatman, N. (ed.) (1994) *Field Margins: Integrating agriculture and conservation*, BCPC Monograph No 58, British Crop Protection Enterprises Ltd; *Figure 7.3* Rew, L. J., Cussans, G. W., Mugglestone, M. A. and Miller, P. C. H. (1996) 'A technique for mapping the spatial-distribution of *Elymus repens*, with estimates of the potential reduction in herbicide usage from patch spraying', *Weed Research*, **36** (4), pp. 283–292, Blackwell Scientific Publications Ltd.

Tables

Table 4.4 Addiscott, T. M., Whitmore, A. P. and Powlson, D. S. (1991) *Farming, Fertilizers and the Nitrate Problem*, CAB International; *Table 5.1* Bradley, R. and Wilesmith, J. W. (1993) 'Epidemiology and control of bovine spongiform encephalopathy (BSE)', *British Medical Bulletin,* **49**(4), 1993, The British Council.

Colour plates

Plate 2.1 Penrose Film Productions, Chiddingly; *Plate 3.1* F. Hoffmann–La Roche AG, Basel; *Plate 3.2* Dr Jeff Woods, Department of Meat Animal Science, University of Bristol; *Plate 4.1* Rowland Hilder, reproduced from *Rowland Hilder's England,* 1986, The Herbert Press.

Index

Note: Entries in **bold** are key terms. Page numbers in *italics* refer to figures, tables and colour plates.

abattoirs, 123
  regulations for BSE control, *58*, 59, 64, 66, 67
accumulated data, **80**, 81, 82, 125
affluence, food quality effects, 24, 31, 112
agricultural commodity prices, 69, 87
Agricultural Development and Advisory Service (ADAS), 40, 47, 98
agricultural landscape, 9, 42, *Pl. 4.2*
agricultural policy, 8, 47, 103
  see also Common Agricultural Policy
agricultural price support, 10, 47
agricultural production, excessive, 8, 47
agricultural science, 5, 6
agricultural surpluses, 90–1, 128
agriculture
  comparative study, *see* LIFE project
  conventional, 6, 89, 92, 97, 128–9
  integrated, *see* integrated agriculture
  organic, 38, 71, 75, 89
  settled, 5
  sustainable, 6, 92
aldrin, 72
algal blooms, 37
  see also eutrophication
Almond, J., 66
amino acids, 14, *21*, 114
  essential, 10, 20
  pig requirements, 20–1, 22
ammonia (NH_3), 38
ammonification, *37*, **39**, 119
ammonium ion (NH_4^+), 38, 39–40, 48, 99
ammonium nitrate fertilizer production, 36, 118
amphidiploids, 106
animal breeding, 26–8
animal manure, *37*, 38
animal production systems, **16**
annual crops, **11**
antibiotics, 23, 31, 107–8
aphidophagous insects, 95, 96–7
aphids, 15, 73–4, 94, 123–4
  predators, *see* aphidophagous insects
aquifer, 32, 48
arable land, conversion to grassland, 47
area farmed, *7*
arsenates, 69
autumn ploughing, 42, *Pl. 4.1*
autumn-sown crops, 11, 42
back-fat (pig-meat), 29, 30
  breeding programme for reduction, 28
  changes with selection, 28
  depth, 25
bacteria, *37*, 38, 39
  denitrifying, 43
barley, 5, 94, 110–11, 113
  in pig diet, *21*, 107
  production, *7*, 9, 113
Barley Yellow Dwarf Virus, *94*, 95, 110
beef, *7*, 50, 66, 67
beef cattle, 53, 122
  BSE in, *54*, *55*, 56
beef industry, 109–10
'beetle banks', 95
benomyl, 75
biennial crops, **11**
bioassay, 58
biochemical/biosynthetic pathways, 12–13, 85
biological pest control, 6, 87, 95
biomagnification, **72**, 85
biomass, **39**, 48, 119
bipyridinium herbicides, **76**
birds of prey, 72
'blue-baby syndrome', 34, 117
  see also methaemoglobinaemia
Bordeaux mixture, 75
bovine spongiform encephalopathy (BSE), **50**–68
  age-specific incidence, *54*, 56, 59, *60*, 122
  causative agent, 61–3
  chemicals not involved, 56
  compensation payments, *58*, 66, 67, 123
  cross-species transmission, 50, 58, 62–3, 67
  risk to humans, 63–7, 123
  direct transmission between cattle, 59–61
  epidemiological investigations, 53–6, 67
  estimated human exposure, 64
  geographical distribution, *55*, 122–3
  government handling of epidemic, 64–6
  identification of infective tissues, 57–8
  incubation period, 56, 58, 65, 122
  measures to control, 57–9, 63–4, 65–6, 123
  numbers of affected cattle, 50–*51*, 58–9, *66*
  possible link with CJD, 50, 59, 63–7, 109–10
  sex-related incidence, 54, 122
  single strain, 63
  symptoms, 51, *52*

Brassica species, 10, 13, 113
breeding policy
  pigs, 26–9, 31
  rape, 10–11, 13, 14, 15
Brundtland, G.H., 91
BSE, *see* bovine spongiform encephalopathy
butter 'mountains', 90
cabbage aphid, 15
cabbage butterfly, 15
cabbage/kale, 10, 106
Calman, K., 66
cancer, 35–6, 118–19
captan, 75
carbamates, 72
carbaryl, 72
carbohydrates, 21
  photosynthetic production, 38
  in pig diet, 21
carboxin, 75
carcass (pig) quality, 29, 30
'catch' crops, 42, 46, 48, 93
cattle, 39, 41–2
  farming systems, 52–3
  see also bovine spongiform encephalopathy
cattle feed
  BSE control measures, *58*, 67
  possible origin of BSE infection, 56–7, 60, 65, 66, 67
causal relationships, 117
cereals, 5, 46, 76
  comparison of cultivation methods, *see* LIFE project
  see also barley; wheat
characteristics (animal), distribution, *27*
chemical industry, nitrogen fixation, *37*, 38
chickens, 16–20
chlorinated aliphatic acids, 76
chromosomes
  number in *Brassica* spp., 10, 11, 106
  number in oilseed rape, 113
chronic wasting disease, *51*
Cincinnati, 24
citrus scale-insect, 87
civilization, 5
CJD, *see* Creutzfeldt–Jakob disease
Cobbett, W., 24
colour (yolk), scale, 17–18, *Pl. 2.1*
commodity production, *7*
Common Agricultural Policy (CAP), 46, 47, 89, 90–1, 92, 103
conservation headlands, 95
contact insecticides, 73, 123, 124
control (experiment), 28, 123
conventional agriculture, 6, **89**, 92, 97, 128–9
  compared with integrated agriculture, *see* LIFE project
cooking oil, 10, 13, 14
corn oil, *12*
cover crops, 42, 93, 96
Creutzfeldt–Jakob disease (CJD), 50, *51*, 62, 67
  monitoring programme, 66
  possible link with BSE, 50, 59, 63–7, 109–10
  types, 51–2
crops
  air movement around, 82
  autumn-sown, 42
  diversification/rotation, 92–3, *100*
  domestication, 5
  nitrate uptake, 46, 93
  pesticide removal from soil, 84
  price support, 10, 90, 91, 92
  species use, 5
  storage, 5
  surplus nitrogen, 46
  yields, 9, 11, 47, 69, 87, 90, 101, *102*, 103, 112
  see also 'catch' crops; cover crops
Cruciferae, 10, 14, 15
cultivation methods, 93–4, 110–11
cypermethrin, 72, *73*
2,4-D, 76
dairy herds, 52–3, 122
  BSE in, *54*, *55*, 56, 67
dalapon, 76
DDT, 70, 72, 73, 88, 110
decomposition, *37*, **39**, 40, 119
denitrification, *37*, **43**–5, 49, 109
derris dust, 71
dicarboximides, 75
2,4-dichlorophenoxyacetic acid (2,4-D), 76
dicotyledenous plants, 76
dieldrin, 72
diet
  carcass quality effects, 29–30
  hens, 18
  pigs, 21–3
  see also nutrients
digestible energy, 21
dinitroanilines, 76
diploid number of chromosomes, **10**, 106
diquat, 76
direct transmission of disease, **59**
  of BSE, 59–60
disease, 15, 23, 46
distribution, frequency, *27*, *86*
diuron, 76
domestication of crops, 5
'double-low' oilseed rape, 15, 114
drainage, 40–1, 84
E. coli, 34
economic optimum, 46
  nitrate application, 46

pig diet, 22
eggs, 16–20
commercial production, 19, 20
consumption, *7*, 16
environmental influences on production, 19–20
laying, 19
yolk colour, 16–19, 114–16, *Pl. 3.1*
eggshell thinning, 72
eicosenic acid, *12*, *13*, 107
electrostatic sprayer, 79–80
endosulfan, 72
environmental factors in egg production, 16, 19
environmental optimum, nitrate application, 46
environmental pollution, 37, 42, 48, 71, 84–5
Environmentally Sensitive Areas scheme, 91
enzymes, 13, 14, 38, 83, 84, 87, 114
epidemiological approach (BSE investigation), **53**–6, 67
epidemiologists, **35**, 118
ergosterol inhibitors, **75**
erucic acid, **10**, 11–14, 107
level in oilseed rape, 13, 14, 113
permissible quantity, 13
synthesis, 12–13
Escherichia coli, 34
essential amino acids, **10**, 20
essential nutrients, 14, 20, 21
esters of glycerol, *see* triglycerides
Ethiopian rape, **10**, 11, 106
ethirimol, 75
European Union (EU, formerly EEC), 10, 13, 15, 34, 117
Common Agricultural Policy (CAP), 46, 47, 89, 90–1, 92, 103
policy on BSE control, 50, 59, 67
eutrophication, **37**, 48, 119, 128
evaporation, 41, 83, 84
'eye muscle', 25
faeces (farm animal), *37*, 39, 41
see also farmyard manure
farm workers, number, *7*, 112
farmers, 40, 45
number, *7*, 112
farming methods, *see* agriculture
farmyard manure, 42
fat
animal, 26, 30, 31
back-fat, 25, 28, 29, 30
carcass, 29
content of meat, 25–30, *Pl. 3.2*
deposition with age, 29
dietary, 21, 29
intermuscular, 25, *Pl. 3.2*
intramuscular, 25, *Pl. 3.2*
marbling, 26, 29
polyunsaturated, 14, 26
saturated, 26, 29, 30, 108–9, 113
unsaturated, 26, 29, 30, 31, 108–9, 113
fatty acids, **11**, 106, 113
polyunsaturated, 30
see also linoleic acid
saturated, 12, 14, 29, 30, 108–9
'short-hand' form, 12
unsaturated, 12, 13, 14, 29, 30, 107, 108–9
fertilizers, *37*, 45, 97–8, 112, 121
addition to soil, 40
application of, 38
BSE control measure, *58*
production, 36, 118
requirements for grazing, 41–2
use in conventional agriculture, 92
see also nitrogen fertilizer
fixation of nitrogen, *37*, **38**, 119
food, 5
colour, 16, 17
prices, 48
visual appeal, 16, 19, 28
food chains, 70, 72, 85
food 'mountains', 8, 90, 128
food production, 5, 8, 16, 40, 47
formulated pesticides, **76**–77
frequency distribution, *27*, *86*, *114–16*
fumigant insecticide, **73**, 124
fungal diseases of crops, 93, 97
fungicides, **69**, 70, 74–5, 88, 97, 101
gametes, 106
GATT (General Agreements on Tariffs and Trade), 89, 91–2, 103
genetic change, 5
genetic engineering, 14, 67, 93
genetic variation, 29
Gertmann–Straussler syndrome, *51*, 62
glasshouse cultivation, pest control, 95
gluconapin, 14
glucosinolate, 10, 14–15, 114
glycerol, **11**, 12, 13, 106, 113
glyceryl tripalmitate, 11, 106
glyphosate, 76
goitre, 15
grain crops, *7*
see also cereals
'grain mountains', 8
grassland, 46–7
conversion of arable to, 47
nitrogen release, 40
ploughing, 40, 46
grazing, 41, 47
greenhouse gases, 45
growth
efficiency, 20

nutrient requirements, 20–1
of pigs, 23
rate, 20

Haber–Bosch process, 38

haemoglobin, foetal, 34

half-life, **83**, 126, 127

haploid number of chromosomes, **10**, 106

hay, *7*, 42

hedgerows, 46, 92

hens, *see* chickens

herbicides, 69, 76, 84, 85, 88, 97, 101
patch spraying, 97, *98*
selectivity, 76

heritability, **26**, 27, 108
experimental assessment, 28

homologous pairs of chromosomes, **10**, 106

hoverflies, aphid predators, 96–7

humus, *37*, **39**, 119
and pesticides, 84

hunting/gathering, 5

hybrids, 11, 113

hydrolysis, 14–15, 84

hydroxypyrimidines, **75**

hypotheses, **53**–4

indirect transmission of disease, **59**

inputs to animal production systems, **16**, 31

insect pests
control in integrated farming systems, 94–7
plant defences, 15
see also biological pest control

insecticides, **69**, 70, 71–4, 84, 85, 86, 88, 101
aphid study, 73–4, 123–4
natural plant extracts, 15, 71
nerve poisons, 73
persistence, 72
routes to target, 72, 73, 124
synthetic, 72
targeted application, 94–5
see also under pesticides

integrated agriculture, **90**, 111, 128–9
comparison with conventional approach, *see* LIFE project
defined, 89
future developments, 102–3
techniques, 92–9
crop diversification, 92–3
disease control, 97
insect control, 94–7
methods of cultivation, 93–4
provision of nutrients, 97–9
weed control, 97

integrated farming systems, 89–103

iodine uptake, 15

iprodione, 75

isotope tagging technique, 42–5, 109, 120

kuru, *51*

late-maturing tissue, **29**

Lawrence, J., 24

laying diets, 18

LD_{50}, *86*, 127–8

leaching, *37*, **39**–46, 48, 93, 109, 118, 119, 120

LEAF (Linking the Environment and Farming), 103

leanness (pigs), 26, *27*
selection for, 28, 29

legumes, 38, 48, 93, 129

LIFE (Less-Intensive Farming and the Environment) project, 99–102, 129

lightning, *37*, 38

linoleic acid, 12, 13, 14, *21*, 106, 107

linolenic acid, 12, 13, 14, 107

livestock diet, 10, 18, 20–3
amino acids, 10, 22, 107
composition for pigs, 20, 21, 22
crops for, *7*
energy-yielding potential, 21, 107
formulation, 22
laying, 18
nutritive value, 22
therapeutic supplements, 23
vitamin supplements, 22

livestock diseases, 23
natural immunity, 23
see also bovine spongiform encephalopathy

livestock feeds, 10, 14, 15, 112
see also cattle feed

lubricants, 14

lysine, *21*, 22, 107

macaque monkeys, spongiform encephalopathy in, 67

mad cow disease, 51
see also bovine spongiform encephalopathy

malathion, 72

manure, 37, 38

marbling, 26, 29, *Pl. 3.2*

margarine production, 13–14

maximum residue levels of pesticides, **71**

MBC generators, **75**

MBM, *see* meat and bone meal

mean, **26**, 80, 81

meat and bone meal (MBM), **56**
possible source of BSE infection, 56–7
prohibited in feeds, *58*

Meat and Livestock Commission, 25, 28

metal-based fungicides, **75**

metalaxyl, 75

methaemoglobinaemia, 34, 36, 48, 117

methionine, 22, 107

mice, experimental BSE transmission, 57–8, 67

microbes, 38, 43, 44, 107–108
in nitrogen fixation, 38

soil, 40–1, 42, 46, 84
see also bacteria
microbial diseases, plant defences, 15
milk, yields, 90
Mills, J., 20
mineral nitrogen (N_{min}), 39–40, 44, 45, 99, 109, 120
mineralization, **39**, 40, 47, 119
minerals, nutrient, *21*, 97–9
minimal cultivation, **94**, 97
Ministry of Agriculture, Fisheries and Food (MAFF), 46, 65–6, 103, 123
monocotyledenous plants, **76**
monoculture, **69**, 87, 92
mustard, 10, 14, 42
mustard rape, **10**, 11, 106
myrosinase, 14, *15*, 114
^{15}N (nitrogen-15 isotope), tagging, 42–5, 109, 120
N_{min}, **39**–40, 44, 45, 99, 109, 120
nasturtium, 14
National Farmers Union, food quality scheme, 103
natural enemies of pests, 72, 87, 88
natural selection, 23, 85, 128
nature-identical synthetic compounds, **18**, 22, 31
nature-related synthetic compounds, **18**, 31
nerve poisons, 73, 88
nicotine, 69, 71
nitrate, 32–49, 119–21
concentrations in rivers, 32, *33*, 37, 42, 48
concentrations in saliva, *36*
control on farm, 40–2
EU limits, 34, 117
in environment, 37, 121
food levels, 119
health problems, 34–6, 117–19
leaching, 39, 40, 42, 43, 44, 109, 118, 119, 120
occupational risks, 36
in soil, 38–40, 41, 42
solubility, 40
surplus, 45–6
untimely, 41, 42, 45, 120
in water, 32, *33*, 34, 35, 37, 42, 47, 117–19
Nitrate Sensitive Areas (NSAs) schemes, **46**–7, 49, 91
nitrification, *37*, **39**, 119
inhibitor, 99
nitriles, 15
nitrite, 34, 35, 36
food levels, 119
formation, *37*, 38
oxidation in soil, *37*, 38, 39
nitrogen
atmospheric, *37*, 38, 39, 43, 45
fixation, *37*, 38
mineral (N_{min}), 39–40, 44, 45, 99,109, 120
nitrogen-15 isotope (^{15}N) tagging, 42–5, 109, 120
organic, *37*, 39, 40, 41
soil, 93, 129
storage locations, *39*
nitrogen cycle, **37**, 39
nitrogen fertilizer, *7*, 32, 39, 109, 112
application, 32, 39, 45, 46, 47, 119
in LIFE project, 100, *101*
profit margins, *98*
reduction of levels, 98–9
excess, 42, 45
food prices, 48
leaching, *see under* nitrate
tagging with ^{15}N, 42–5, 109, 120
time of application, 120
nitrogen oxides, *37*, 38, 119
see also nitrous oxide
nitrogen-fixing microbes, 38
nitrogenase, 38
N-nitrosamine, 35
nitrous oxide, 43, 45
nomadic-pastoralism, 5
normal distribution, **26**, *27*, *86*, 114, 127
number mean diameter, **80**–82, 125
nutrients
essential, 14, 20, 21
pig requirements, 20–1, 22
plant, 73
provision, 16
oil
edible, 10, 12, 13, 14
industrial uses of rapeseed, 14
pesticide diluting agent, 76, 79, 80
oilseed rape, *7*, 9–15, 42, *Pl. 2.1*
area devoted to crop, *9*
biosynthetic pathways, 12–13, 107
double-low varieties, 15
erucic acid content, 11–14
genotypes, 14, 106
glucosinolate content, 14–15
harvest size, *9*
harvesting, 11
industrial uses, 14
low-erucic-acid varieties, 13
production, *9*, 112, 113
sowing, 11
oleic acid, 12, *13*, 14, 106, 107
organic agriculture, 38, 71, 75, 89
organic matter, 38, 39, 40, 44
organic nitrogen, *37*, **39**, 40, 41
organochlorines, **72**, 84, 85, 110
organoleptic characteristics, **19**
pig-meat, 24, 25, 26, 28, 30, 108
organophosphates, **72**, **76**, 110
possible link with BSE infection, 57
outputs of animal production systems, **16**, 31
ovulation, 16, 19, 31

oxathiins, **75**
oxides of nitrogen, *37*, 38, 119
see also nitrous oxide
palm oil, 10
palmitic acid, 11, 12, 13, 14, 106
paraquat, 76
pathogenic organism control, 15, 23, 86
peanuts, 10, 29
persistence of pesticides, **71**, 72, 85
pesticides, 15, **69**–88, 128
adsorption, 83, 84, 85
in air, 84
application, 76–83, 85, 110
depositing on plant, 74, 77–80, 82, 83, *84*
drop size, 77–8, 80, 81, 110, 124–5
proportion on crop, 77
spraying, 69, 76–84, 124–5
targeted, 94–5
breakdown products, 70, 71
carriers, 76–7
discovery and development, 69–72, 75–6
environmental impact, 69, 70, 84–5, 87, 110
formulation, 76–7
harm to natural enemies, 87
human safety, 70, 71, 73, 85
microbial degradation, 84, 85
oil diluting agent, 76, 79, 80
on/in plants, 83–4
in organisms, 85–7
possible hazard to cattle, 57
registration, 71
residue levels, 71, 84–5
resistance to, 72, 74, 75, 85–7, 87, 110, 128
risk assessment, 85
safety data, 70, 71, 85
in soil, 72, 76, 84
testing, 70–1, 87
toxicity, 69, 71, 72, 83, 85
use in conventional agriculture, 92, 100, *101*
use in integrated agriculture, *101*
water contamination, 84–5
see also fungicides; herbicides; insecticides
pests, 69, 87
control
insect pests in integrated farming systems, 94–7
see also biological control of pests; pesticides
predators of, 72, 87, 93, 94, 95–7, 111
propagules, 90
petrol engine, combustion in, 38
Phacelia tanacetifolia, 96
phenoxyacetic acids, **76**
phenylamides, **75**
phloem, **73**, 123, 124
phosphate, role in eutrophication, 37, 48
photochemical oxidation, **83**, 84, 85
photoperiod, **19**–20, 31, 107
phthalimides, **75**
pig production
diet composition, 10, 20–2
disease, 23
disease-free stock, 23
growth, 20–1, 23
manure, 47
meat quality, *see* pig-meat quality
muscle growth, 20–1, 25, 29
nutrient provision, 20–3, 107
'performance', 23
pig-meat consumption, *7*, 16, 20
pig-meat quality, 23–31
breeding policy, 24–5, 26–9
fat content, 25–6, 28–9, 30, *Pl. 3.2*
leanness, 25, 26, 27, 28, 29
measurement, 25–6
modifying, 26–30
organoleptic assessment, 24, 25, 28, 30
processing, 16, 24, 31
visual appraisal, 25, 26, 28, 30, 31, *Pl. 3.2*
pigmenting agents, **16**–17
natural, 18–19
plant breeding, 5, 10, 13, 14, 15, 113
plant defence mechanisms, 15
plant oils, 10, 11–14, 26
ploughing, 42, 93–4, *Pl. 4.1*
pollution, *see* environmental pollution
polyphagous insects, 95
polyunsaturates, 14, 26, 30
population growth, 6, 7
Porkopolis (Cincinnati), 24
poultry, *7*, 10, 16–20, 47
predators
pesticide harm to, 70, 87
of pests, 72, 87, 93, 94, 95–7, 111
prion diseases, *see* transmissible spongiform encephalopathy diseases
prion protein, 51, **61**
cross-species transmission, 62–3
properties, 62
probability, 35, *36*
progoitrin, 14, 15
propagules of pests, **92**
protease resistant protein (PrP), **62**
proteases, ***61***
protectant fungicides, **74**, 75
protein, 10, 14, 20
dietary, 20–1, 22, 107
of rapeseed meal, 10, 14, 113
synthesis, 20
see also meat and bone meal; prion protein
proteinaceous infectious particle, **61**
see also prion protein
PrP, **62**

see also prion protein
PrP^cell^, **62**, 63
PrP^prion^, **62**, 63
Prusiner, S., 61
pyrethroids, synthetic, 72–3
pyrethrum, 71, 72
rainfall, 40–1, 42, 46, 83, 84, 85
rape, **10**, 11, 13, 15
see also oilseed rape
rapeseed meal (cake), **10**
palatability, 14
toxic glucosinolates, 14–15
rapeseed oil, 12, 14
raw material use, 6
reduced cultivation, **94**
rendering, **56**, *57*, 67
replicates (experiment), 73, 123
residual insecticides, 72, **73**, 123, 124
resistance to antibiotics, 23, 107–108
resistance to pesticides, 72, 74, 75, 85–87, 110, 128
resource use, *7*, 19
Ridley, H. N., 5
risk, 34, 35, 36, 69, 85
risk/benefit analysis, **69**
rivers, nitrate concentrations, 32, *33*, 37, 48, 117
root crops, *7*
rotenone, 71
Rothamsted Experimental Station, 40, 43–6, 48, 120
route to target, 72, **73**, 88, 123–4
Royal Shows, 25
rubber crop, 5
saliva, 35, *36*
sample size, 116
SBOs, *see* specified bovine offals
scrapie, *51*, **52**, 123
causative agent, 61
see also prion protein
cross-species transmission, 62–3, 65, 66
possible transmission in MBM, 56–7, 67
strain variation, 63
seed stocks, pure, 5
selection
for leanness (low-fat), 27, 29
natural, 23, 85, 128
response to, 27
selection pressure, **27**, 110
set-aside land, 91, 100
settled agriculture, 5
silage, 42, 112
simazine, 76
Simmonds, N. W., 5
slash-and-burn arable farming, 5
slurry, 42
soil, *37*, 38–40, 42
drainage, 40–1
erosion rosk, 93, 94
fertilizer nitrate loss, 46
microbes, 40–1, 42, 46, 48, 70, 84
nitrate in, 38–40, 99
nitrate leaching, 46, 93
nitrogen, 40, 93, 129
pesticide in, 72, 76, 84
type, 46
somatic cells, 10, 106
Southwood, R., 65–6, 123
soya bean, 21
soya bean oil, 10, *12*
species barrier, 50, **62**–3, 67
specified bovine offals (SBOs) or materials (SBMs), **58**, 63–4, 67, 123
spinning cage nozzle, 78, 83
spinning cup nozzle, 79, 80, *81*, 82
spray drift, 77–8, 84
spraying, 69, 76–84, 125, *126*
drop size, 77–83, 125, *126*
electrostatic sprayer, 79–80, 88
equipment settings, 80–3
spinning cage nozzle, 78, 83, 88
spinning cup nozzle, 79, 80, *81*, 82, 88
standard deviation, **27**, 108
statistical significance, 35, *36*, 118
stearic acid, *13*, 107
stocking density, 23, 41
stomach cancer, 35–6, 48, 118
subclinical disease condition, **23**
sub-lethal effects of pesticides, 72
subsidies, 90–1, 92
sucklers, *see* beef cattle
sulphur, 14, 69, 75
sulphur-based fungicides, **75**
sunflower oil, 10, *12*, 14
support prices, 90, 91, 92
sustainable agriculture, 6, 92
see also organic agriculture
sustainable development, **91**, 103
symbiotic associations, 38
synthetic pyrethroids, **72**–3
systemic fungicides, **74**, 75
systemic insecticides, 72, **73**, 123, 124
TCA, 76
therapeutic agents, 23, 31, 107
thiocyanate, 15
thyroxin, 15
Townsend, 'Turnip', 5
toxicity

pesticide, 69, 71, 72, 83, 85
rapeseed, 15

tractors, number, *7*, 112

translaminar insecticides, **73**, 123, 124

translocation, **73**, 76

transmissible spongiform encephalopathy (**TSE**) diseases, **51**, 67
cross-species transmission, 62–3
inherited types, 62
in zoo animals, 58

triadimefon, 75

triazines,

trichloroacetic acid (TCA), 76

trifluralin, 76

triglycerides, **11**, 12, 13, 14, 113

Tropaeolum majus, 14

turnip-rape, **10**, 11, 106

ultraviolet light, 83, 85
effect on scrapie agent, *61*

United Nations Conference on Environment and Development (1992), 91

untimely nitrate, **41**, 42, 45

urea, in urine, 41

ureas, herbicides, **76**

urine, *37*, 41, 42

variability of features, **26**, 27

variance, **27**

viruses, 61
diseases in crops, 94, 95

vitamins, *21*
supplements, *21*, 22
vitamin C, 22, 118
vitamin F, *see* linoleic acid

volume mean diameter, **80**–82, 125

water
extraction, 48
nitrate in, 32, 34
pesticide contamination, 84–5
selling, 48
supply of clean, 32
transpired, 84
'well-water methaemoglobinaemia', 34

weather, 46, 77

weedkillers, *see* herbicides

weeds
control in integrated farming systems, 97, 101
pesticide removal from soil, 84
see also herbicides

wheat, 5, *7*, 11, 21, 42, 43–6, 47, 113
control of aphid infestation, 96
control of fungal disease, 97
in conventional agriculture, 92
yields, 90, 101, *102*, 128–9

whitefly control, 95

wild flowers in arable fields, 95–6

wild oat, 76

Will, R., 66

woodland, 46, 47

xylem, **73**, 123, 124

yolk colour, 16–19, 114–16

yolk colour fan, 17, *Pl. 3.1*